SECRETS OF THE HUMAN BRAIN

SECRETS OF THE HUMAN BRAIN

DR. ROBERT PULLEN

YOUCAXTON PUBLICATIONS

ISBN 978-191117-56-50
Printed and bound in Great Britain.
Published by YouCaxton Publications 2017

YouCaxton Publications
enquiries@youcaxton.co.uk

Contents

ABOUT THE AUTHOR

Dr. Bob Pullen is a University Lecturer who studied neuroscience as a student at London and Oxford University. He has spent his working career undertaking research in neuroscience and teaching university students of Medicine, Pharmacy and Biomedical Sciences. His research has investigated Stroke, Hydrocephalus and Dementia. This was undertaken at the Institute of Neurology in London, Brown University in the USA, the Medical School at Newcastle, the School of Pharmacy at the University of Sunderland and the Medical Research Council Dementia Laboratories in Newcastle. He has published many articles about neuroscience in Scientific Journals.

CHAPTER 1

What's My Brain Made Of?
(Brain Cells)

Synopsis: This Chapter explains how life and the brain are made of cells which have evolved with time.

Cells

The human brain is the 'Command and Control' centre of the body and within it are our emotions, thoughts, mind, personality, intellect and self. It is well established that humans have the most complex brains on Earth, but it is possible that creatures with bigger and better brains than ours live elsewhere in the universe. However, we have no evidence as yet that there is anyone else 'out there' and if even they were, they may not be built like us, so for the time being we have the most complex brains in the entire Universe!

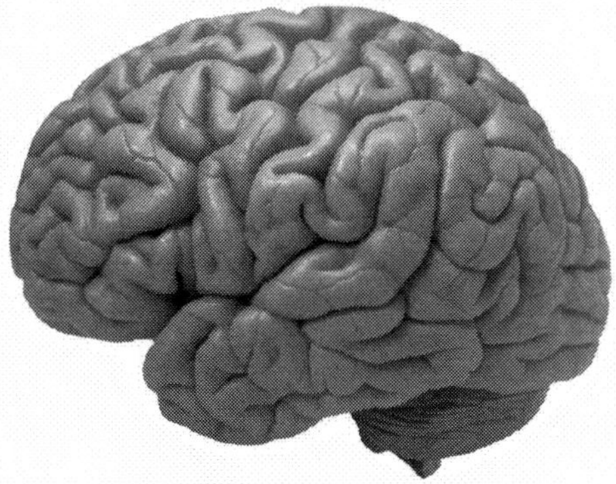

The Human Brain

Our brains are made of cells just like the rest of the human body. Cells are the building blocks of life which are too small to see, so you could place a couple of thousand of them on a pin-head. A few cells are exceptionally large and these are some special nerve cells found in the human brain which are nearly one meter long! All human cells, including brain cells are small bags of salt water which are the building blocks of tissues, creatures, life and us. Cells probably evolved in the oceans where all the required components to build a cell eventually met **(1).** All our cells contain even smaller, sub-cellular structures or units which have special functions. One example of a sub-cellular structure is a unit called a mitochondria. This structure provides energy to the cell to keep it alive. To use the analogy of a car; if the cell is the whole car, the mitochondria would be the engine.

It is thought that just as life, creatures and humans have evolved on Earth, so have cells. A long time ago, probably a few billion years ago, there were cells floating around in the oceans that did not have mitochondria inside them. They were like cars floating around the seas without engines. At the same time, there were also mitochondria (car engines) floating around in the same oceans without cars. One day, by chance a mitochondria collided with a cell. That is, a car engine collided with a car and the two of them together worked a lot better than each alone. And so the modern cell was born. Although you may think this meeting was a very unlikely event. It was. You might say that it would take a million years before the two met by chance and perhaps it did. But remember, there has been plenty of time available for this chance meeting to happen and unlikely things do occur. It is extremely unlikely that someone will win the lottery each week, but someone usually does. The inclusion of mitochondria in a cell is part of cell evolution and this process has not finished. It is likely that one day a modern human cell will collide with some new structure or entity called a 'Z-factor'. The Z-factor may be floating around in space somewhere or is in the air on a distant planet and when the two meet, a new super-improved cell will be born and human cells and life will evolve and change again.

Our cells today as mentioned, are tiny bags of salt water and you may wonder why cells don't have pure water inside them. Why bother to have salt there? One answer is that many of the chemical processes that take place inside cells to keep them and us alive, require salt to be present. This answer, might lead you to also ask 'But why did cells 'choose' common salt, that is sodium chloride rather than something else? The answer is; imagine yourself trapped in a wood and you want to build a shelter there. What might you use? Well look around and of course you see trees everywhere, so it makes sense that you make your house with wood. Similarly, if you were in a stone quarry you would probably choose to build your house with stone. It is likely that cells were first made in the oceans so if you follow this logic and if you want to build a living cell and you are surrounded by salt water, you build your cell using salt water. Had our oceans been full of potassium chloride instead, I dare say our cells would have been built around potassium chloride rather than sodium chloride. Had this been the case, the resulting cells and subsequent life would have been rather different. It would be 'Life Jim, but not as we know it'!

Who Knew What About Cells?

Just about everyone today knows that plants, creatures, life, us and our brains are made of cells, but this has not always been common knowledge. Two thousand years ago, although the Romans were a rather cleaver and sophisticated bunch of folk, they did not have the faintest notion that humans were made of cells. In fact much of biology and science was a complete mystery to them and the rest of the world at that time. Given the limited understanding of human biology at that time, it's not too surprising that past civilisations didn't really know how to kill people apart from sticking a sword through them or chopping their heads off; which did work rather well!

Since the time of the Romans, it took about another thousand years before science progressed to discover cells. This happened in the 1600s with the invention of the microscope which allowed biologists to actually 'see' a cell **(2)**. But even though human cells have been known to exist since then, the structure of the cell wall that encloses the salt water has only been understood in the last hundred years. The human cell wall, including the human brain cell wall, was thought to be a pretty simple, rigid structure until 1935 when two scientists Hugh Davson and James Danielli published some ground-breaking research in which they showed that cell walls were complicated, made of two layers of lipid (fat) they were bendable and had all sorts of complex structures embedded in them **(3)**. In the 1970s this description was built upon to become the 'fluid mosaic model' of the cell wall and this model still remains to-day, as shown **(4)**.

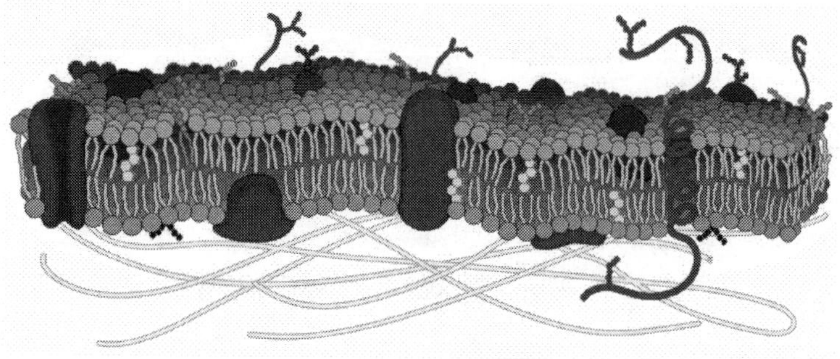

THE CELL WALL

Hugh Davson was an English scientist who died in 1996 aged 87. He married the society portrait painter Marjorie Heath with whom he had one daughter and he was a cousin of the journalist and broadcaster, the late Alistair Cooke. He became interested in brain science and he wrote a famous book called 'The physiology of the cerebrospinal fluid' first published in 1967 **(5)**. Later on, he particularly became interested in the human brain which is also my area of study. In the 1980s Hugh Davson was working at Kings College, London as a Professor and I was fortunate enough to study with him there and discuss research and brain science. He liked his beer and he always bought me a pint in the 'Marquess of Anglesey' in Covent Garden. He was a very bright and kind man and very humble with it.

References and Further Reading:

1. Berne & Levy. Physiology (6th Edition, 2010)
   Koeppen & Stanton
   Elsevier

2. The Evolution of the Microscope (2014)
   Bradbury
   Elsevier

3. Danielli & Davson (1935).
   A contribution to the theory of permeability of thin films.
   Journal of Cellular and Comparative Physiology. 5 (4): 495–508.

4. Singer & Nicolson (1972).
   The fluid mosaic model of the structure of cell membranes.
   Science 175 : 720-31

5. Physiology of the Cerebrospinal Fluid (1967)
   H. Davson
   J & A Churchill

CHAPTER 2

How The Brain Grows

(Brain Growth & Development)

Synopsis: This Chapter explains how the human brain grows and develops and how it has old and new parts. It explains how brain cells are programmed to live and die and the involvement of DNA, genetics and evolution. It discusses how our bodies change over time.

Human Cells

Cells are the building blocks of life and every one of them, including brain cells start out in the developing embryo as simple, standard, non-specialised cell. By analogy with cars, all cars start out with a basic frame, a body, four wheels and an engine and these can be specialised into an f1 racing car, a family saloon or a 4 x 4. So it is with cells. They all start out simple and then they become specialised into kidney cells, skin cells or brain cells etc. This specialisation is pre-determined by the cell's DNA in the embryo.

How to Build a Brain

The brain is constructed during the complex process of building a human, which takes about forty weeks; i.e. pregnancy. Biologists have known for many years that every cell in the human body contains an 'Instruction Manual' that explains how to build a complete human from start to finish, although they didn't know where this manual was or how it worked. In the 1950s two biologists working at Cambridge University named Watson & Crick discovered that the Instruction Manual was a special molecule called DeoxyriboNucleic Acid or DNA for short. They went on to describe how the molecule works and they were awarded the Nobel Prize for this in 1962 **(6).**

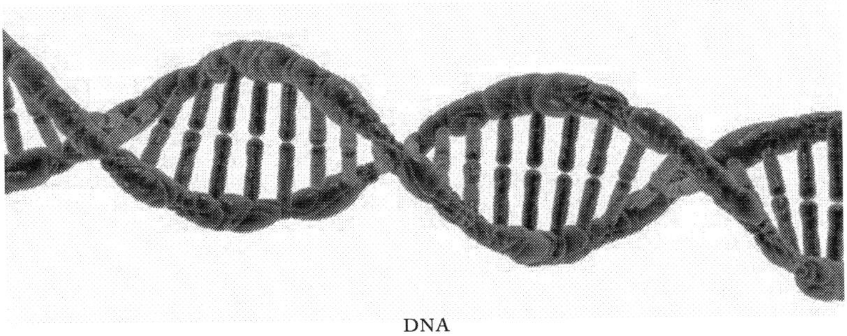

DNA

DNA has a structure rather like this. Imagine you have a metal chain about one meter long which is made of twenty metal links joined together. You then get a second metal chain which is identical to the first and you hold both of them out in front of you side-by-side so they hang down towards the ground. You then twist them around each other. This double-chain is rather like the structure of DNA and chemists call this type of molecule a double-helix. In my metal chain all the links are the same but in DNA there are four possible different links that can be used called A, T, G and C. In DNA the links are particular chemicals and when the builder of the DNA molecule starts to construct the chain he / she can choose to use any of the four possible links in any order they wish. For example, the maker may use three As, followed by six Cs, followed by eight more As, followed by one T, followed by another A followed by one G to complete the chain. In my metal chain I have used twenty links but in real human DNA there are about three billion links!

Three billion links in a row makes a very long DNA molecule and nature has decided to chop it into forty-six bits to make it easier to use and store. These forty-six bits are called chromosomes which are stored inside the cell. In my metal chain it doesn't really matter what order I join the links together because they are all the same, but in real DNA the sequence of the links of A, T, G or C is crucial. It is crucial because the order or sequence of the links has been found to be the code that becomes the 'Instruction Manual' in the cell which explains how to build a human. You could think of the forty-six chromosomes as being forty-six chapters of the Instruction Manual and actually, each chapter has many sections. These smaller sections describe how to make one small part of a human and these are called genes. Humans have about 25,000 genes.

Could There Be Two Of Me?

Humans are normally constructed by a sperm fertilising an egg, but since each cell contains a full set of instructions of how to build a person, geneticists realised that if they could read the instruction manual themselves they might be able to kick-start a cell to grow into a complete person. This process is called cell 'cloning' and in 2003 some geneticists in Scotland managed to 'clone' a whole sheep by kick-starting growth in a cell taken from its' mammary gland. The 'cloned' sheep which was genetically identical to its 'mother' lived for about six years. The scientists named it Dolly, after Dolly Parton. Other animals have since been cloned and in theory, cloning could be applied to humans which raises many ethical and moral questions.

A Brain Like Swiss Cheese!

When it is decided to build a human, the DNA instructs special cells of the foetus to become the brain. These cells rapidly divide and start to form a shape like a tube, a bit like a water-pipe. The walls of this pipe become the brain itself and the hole in the centre of the pipe fills with salt water. The cells of the developing brain rapidly divide at the astonishing rate of about 250,000 new cells (neurons) every minute and this continues for the first three months of pregnancy, the first trimester. After this period, brain cell division stops and it never starts again!

As these brain cells divide the 'tube' grows larger and it eventually becomes the developed brain. The 'tube' (brain) is pulled, pushed and twisted around as it grows and this means that the tube's central water-filled hole also becomes pulled and twisted around. This hole ends up as four fluid-filled chambers inside the brain called ventricles.

These ventricles are irregular shapes and this means that the adult human brain is not a solid 'lump' of cells but rather it has these fluid-filled 'chambers' inside it, a little like Swiss cheese. The reason the tube is pulled and twisted around is because during the first three months of foetal development, the brain speeds through its entire evolutionary history thereby changing shape as it grows. This means that the brain dashes through about 300,000 years of it's evolutionary history each day for about one hundred days!

Modern brain science is not sure why these four fluid-filled caverns or ventricles are present inside the brain. It could be because the brain is rather jelly-like so they provide buoyancy and also they provide some space for the brain to swell into, if it had to! Science is still not sure where all the fluid that fills these chambers comes from or where it goes to? This fluid is called the Cerebro-spinal fluid and sadly, if a baby is born with problems with this fluid it can cause the brain ventricles to expand and crush brain tissue producing a condition called hydrocephalus, meaning water-on-the-brain **(7).**

Some of my research investigated where this fluid comes from and drains to with the hope that it might advance the treatment of Hydrocephalus. I was lucky enough to do much of this research in the Neuroscience Laboratory of Professor, Sir Colin Blakemore at Oxford University **(8).** Colin was also a neuroscientist and became Head of the Medical Research Council and we often met in the laboratory to discuss our research. He was especially interested to know how our brain manages to take the two-dimensional images from our eyes and fuse them into a single three-dimensional 'motion picture' called sight, which happens in the 'blink of an eye'. And we still don't know how it all happens! See later.

The last change in the brain growth is the appearance of the two large frontal lobes and their arrival has meant that our skull has had to enlarge to accommodate them. Early humans had a swept-back forehead reducing the volume of the cranium so with the evolutionary growth of the frontal lobes modern humans have a vertical forehead above the eyes which expands the front of the cranium to accommodate them **(9)**. This particular change has had huge consequences in the development of speech and language, see later.

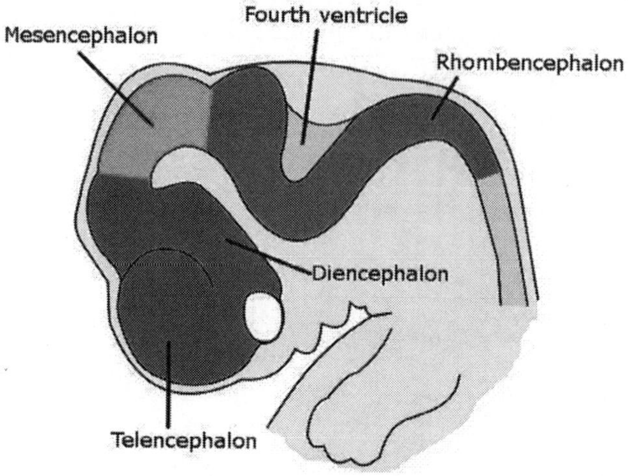

BRAIN DEVELOPMENT

Have We Evolved?

Having discovered DNA and the code it holds, biologists set out to read the sequence of all the links of the DNA chain that is, the sequence of all the As, Ts, Gs and Cs in one set of forty-six human chromosomes. This required reading all three billion 'links' and this was achieved in 2001. Since then, many more sets of human chromosomal DNA have been read & sequenced and also the DNA of many other animals and bacteria. This research has shown that the sequence of the links in the DNA of primates like chimpanzees and gorillas is more than ninety-five percent the same as humans. Other recent genetic research has found that many genes found in human DNA are also found in many other animals, plants, fungi and even bacteria. Taken together, these observations suggest a common origin of cells and life which is used by supporters of Darwinian Evolution **(9)**.

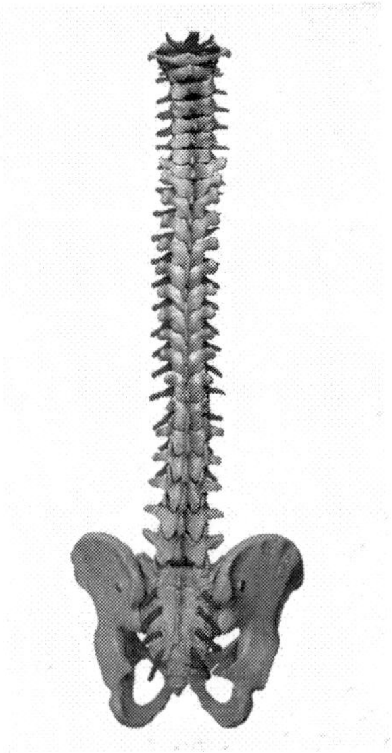

Our Segmented Spinal Column

Not only does the human brain trace it's developmental history as it grows in the foetus, so does the whole human body. Early creatures on Earth were constructed of repeating units or segments so they were 'segmented'. This can be seen as far back as earthworms and it continues into fish and more recent vertebrates like cats, dogs and us. Our body's final segmentation can be seen by looking at the final thirty-three vertebrae of our spinal column **(1).**

Time To Get Changed!

Humans and human cells don't live forever. All our cells have a pre-determined life-span and the duration of this is different for different cell types. For example, blood cells live for about three months, surface-cells of the skin live for about one month whereas stomach-lining cells only live for a few days. However, just before a cell dies, it divides and reproduces itself and the 'off-spring' cell then replaces the dying parent cell. This means that the cells of the human body are in a continual state of birth, division, death and replacement or you could say in a state of continual 'turn-over'. For example, our blood is replaced in total about every three years and the lining of the stomach is replaced in total about every month. As an aside, this does ask the question 'Why don't skin tattoos disappear from the skin if it is replaced every month? The answer is that much of the ink in a tattoo is held in cells below the surface of the skin and when these cells die, they pass on their contained ink to the new replacement cells.

There are more changes taking place in our brains and bodies which are superimposed upon these cellular changes. These are molecular changes because the atoms and molecules that make up the cells are also constantly changing. By analogy, think of a liver cell to be like a house made of bricks. The whole house is one liver cell and the house bricks are the molecules that make up the cell. Suppose that one year the house owner replaced the bricks of one wall of the house with new bricks. At the end of the year you would say it is the same house even though it has a new wall. If the next year they replaced the bricks of a second wall with new bricks, at the end of the second year you would probably still say it was the same house, but with two new walls, even though it would be different from the original house. If they repeated this process several times, you would eventually reach a point where all the bricks of the original house had been replaced by new ones. At that point you could argue it had become a new house, but you would probably still say it was the same house, even though none of the original house bricks were present. In the same way, the molecules that make up our cells are constantly changing and being replaced and it has been determined that all the molecules that make up all our cells that is, our whole body, are changed about every twenty years **(10).**

Drink Anyone?

If this sounds odd to you, it might help to think about our body and water. Our cells are largely bags of salty water and the water inside them is constantly moving and mixing with the water that is outside them which is constantly mixing with blood etc. Each day our body loses about two liters of water and we replace this by drinking about two liters of fluid each day. This Illustrates that there is a constant, daily turn-over of water in our body and over months this means that all our body water is eventually replaced. The same applies to all molecules in the body. As an interesting aside, water is found all over the Earth with billions and billions of water molecules present in the oceans, clouds and in the tap water in our houses. If all these water molecules across the whole planet were being constantly and perfectly mixed and you drew yourself a glass of water from the tap in your house, there is a 99.9% chance that one molecule of water in your glass was once in a tear shed by the Egyptian Boy King Tutankhamun, or any other person you choose from the past!

TUTANKHAMUN

That's All You Get!

I have mentioned that different cell types of the human body have different turn-over times but there is one important exception and that is the brain. After fertilisation, all other human cell types in the growing foetus continue to grow and divide throughout life, but adult brain cells don't divide. This means that once the foetus is three months old it will never grow any more brain cells which means that the total number of brain cells we have at that point in pregnancy is the maximum number we will have for the rest of our lives! This surprises many people who remark that a baby's head certainly grows and enlarges after birth so that an adult's head is much larger than that of a child. This is true and the brain does enlarge inside the head as the cranium grows, but this is not due to the growth of new neurons. It is due to several other events. One is the growth of another type of cell found in the brain called glia cells. A second factor is that many of the neurons of the new-born brain are rather compressed together and in early life these spread apart occupying more volume. A third reason for brain enlargement is that all the young neurons start to grow new connections between themselves and this new growth takes up space.

Brain cells start to die after birth so the total number of neurons in our brain slowly reduces as we get older. Normally, we have enough brain cells at birth to keep the brain working properly for about one hundred and fifty years, so humans could live that long. However, nature has determined that some other part of the human body deteriorates before then so most of us live for about seventy years.

Do You Really Change?

Because adult brain cells can't divide and reproduce, the brain cells we have at birth stay with us for life which means that our brain cells have a life-span which is the same as our body that is, also about seventy years. This does ask the question 'Why don't brain cells divide and reproduce like all other cells do?' One answer is that if our brain cells 'turned-over' every few months, then our individual selves, our own mind and 'ourselves' might also 'turn-over' with them so that each of us would be a constantly changing person and this might be just too much for the human mind to cope with! This subject then becomes more philosophical than physiological so best leave it there!

Although brain cells (neurons) can't divide and replace themselves, they can make and break the connections they have with each other and during the first years of life there is a huge increase in neuronal connections as the brain's total network grows. After the age of about five years the growth of the network slows but it never stops. Throughout our lives, our neurons are constantly making new connections and breaking others. It is thought that this process is part of learning and repeating a thought or task reinforces connections so that 'practice makes perfect '!

The fact that our brain cells can't divide has other consequences for us. In all other tissues, when cells die or are damaged, new cells are made which replace the damaged ones and the tissue is repaired. Since brain cells can't divide this repair process cannot take place making it vulnerable to damage and that is one reason why the brain is encased in a hard, protecting skull! Another consequence of brain cells not dividing is that as we lose brain cells throughout our lives there will come a time when there will not be enough cells present in the brain for it to work properly. Something else may kill us before we reach this point, but sometimes a significant loss of brain cells can be observed in life and this is commonly seen as dementia. And because brain cells are continually dying throughout life, dementia always progresses and can't be cured, see later.

Eternal Life Please

I have mentioned that human cells and humans don't live forever. All human cells whatever type (except brain cells) grow, live, divide and then die. This limited life-span is programmed into a cell's DNA and there are complicated processes that control it called apoptosis **(10).** Sometimes apoptosis goes wrong and cells start to grow and divide rapidly and in an un-controlled way which can lead to 'cancer'.

In 1951 there was an American patient called Henrietta Lacks who had cervical cancer and some researchers took cells from her cancerous cervical tumor and she died later that year. In the laboratory the scientists found her tumour cells were remarkably durable and they spent much time trying to keep them alive. As mentioned, normal cells don't live forever, but the scientists managed to keep these abnormal cancer cells alive and dividing in the laboratory for years. In 1955, these cells were 'cloned' and it was found that the new cloned cells never died because the process that ages them and causes their death, apopotisis, had stopped working. These cells were called 'HeLa' cells, chosen from Henrietta's name and they were immortal! This immortality produced a great demand for these cells from scientists around the world who wanted to study them in their laboratories and to date, about twenty tons of her cells have been grown and sold!

HeLa cells are not normal in a number of ways. Firstly, they are cancerous but also they don't have many of the required properties of normal working cells, such as they can't be specialised into a particular cell type. These differences have led some experts to believe that these HeLa cells are not human and could be a new 'species'. At present, scientists can not properly control or manipulate the growth of HeLa cells but in the future when genetic engineering has become more sophisticated, this should be possible. At that point, the control of the growth and death of HeLa cells could also be applied to all 'normal' cells that comprise the whole human body and if this happened we would have eternal life, which would lead to a rather crowded world!

References and Further Reading:

6. The Double Helix (2010)
Watson
W & N

7. Gray's Anatomy (41st Edition, 2015)
The Anatomical Basis of Clinical Practice
Stradring
Elsevier

8. Pullen, R (1985).
Effect of ventricular tonicity on CSF production.
Journal of Physiology 362, 273-283.

9. The Ancestor's Tale (2016)
Dawkins & Wong
Clays

10. Principles of Neural Science (5th Edition, 2012)
Kandel, Schwartz, Jessell, Siegelbaum & Hudspeth.
Elsevier

CHAPTER 3

Bright Sparks
(Types of Brain Cells)

Synopsis: This Chapter explains how the brain is composed of nerve cells (neurons) and it outlines how these behave like batteries and like computer-chips. It explains how the brain has its own language which is like 'Morse Code'.

Neurons

There are two types of cells in the human brain, nerve cells also called neurons and glia cells. Together these two types of cells comprise the whole brain which weighs about one kilogram. The Glia cells in our brains are much larger than neurons and make up about sixty percent of the mass of the brain. They are not well understood but are thought to support neurons in various ways.

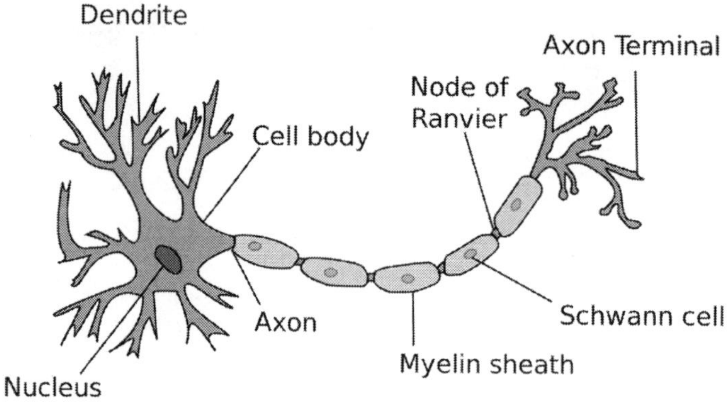

Brain Neuron

The nerve cells or neurons are the main cells of the brain. Each neuron has finger-like projections the axon terminals, that connect to many other neurons. Typically, a single neuron will send out about 5,000 connections to other neurons and the output signals from the neuron pass down these connections to these cells. The same single neuron also receives input connections from other neurons on special regions called dendrites. Typically, a single neuron will receive about 20,000 connections (inputs) from other brain neurons. So, a single brain neuron is part of a very extensive neural network with about 20,000 inputs and 5,000 outputs. Furthermore, the brain has many, many neurons and since each has its own extensive network, the network of the whole brain is enormous. Think of it this way: as mentioned, cells are very small, but if you enlarged a brain neuron to be the size of a can of Coca-Cola and you had a can of Coca-Cola for every neuron you have in your brain and you stood all these cans of coke side-by-side, they would cover the entire continents of the Earth... and all the oceans. Yes, they would cover the entire surface of the planet. Now that is some network and connectivity!

Neurons behaving like Batteries

Neurons have two particular functions. Firstly, they can behave like a 'battery' and produce a voltage and secondly, they can also behave like a 'computer-chip' and process electronic signals. When a neuron is in 'battery mode' it is able to manipulate the electrically-charged sodium ions (and potassium ions) it has contained in its water and it can manufacture a small electrical 'spark' which is about one tenth of a volt. This voltage or spark is called the 'Action Potential' which lasts for a fraction of a second and when required, a neuron can manufacture about 200 sparks every second.

Neuroscience has known for years that brain neurons produce these sparks of electricity but this has not always been known. For a start, electricity was not really discovered until the 1700s so before then no one could possibly know that neurons could manufacture a voltage **(11)**. Of course, at the same time, no one new neurons existed either! It was not until the 1800s that scientists worked out that the human brain had neurons which produced sparks **(12)** and it was not until the 1950s that scientists worked out how the neuron is able to produce this 'spark.' To study the neuron action potential, neuroscientists wanted to apply electrodes to the neuron to electrically stimulate and recorded from it. However, human neurons are far too small to allow this. By chance, some of these neuroscientists heard that the invertebrate the 'Atlantic Squid' has a giant nerve in its body which it uses to control it's jet-propulsion. This nerve is several centimeters long and about one millimeter in diameter which makes it huge on the biological scale. Importantly, this squid neuron is little different from a human neuron, except it is thousands of times larger. So, this giant nerve was extracted from the squid and it was big enough for neuroscientists to use electrodes on it. In result, two neuroscientists named Hodgkin & Huxley, discovered how the spark is manufactured and they were awarded the Nobel Prize for this work in 1963 **(13)**.

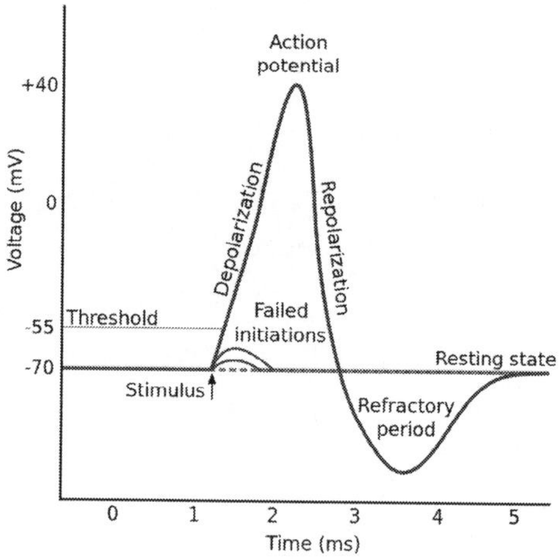

NEURON ACTION POTENTIAL

All 'sparks' made by all brain neurons are the same and it is these sparks that brain neurons send to each other down their connections and throughout their networks. Although all these sparks (Action

Potentials) are identical, modern neuroscience has also worked out how these sparks become signals that become messages to other brain neurons, see later. Once this was established, it was entirely logical for the scientists of the day to assume that brain neurons were 'hard-wired' together by their connections much like electric wires in a building are connected together by junction boxes and plugs fit directly into sockets.

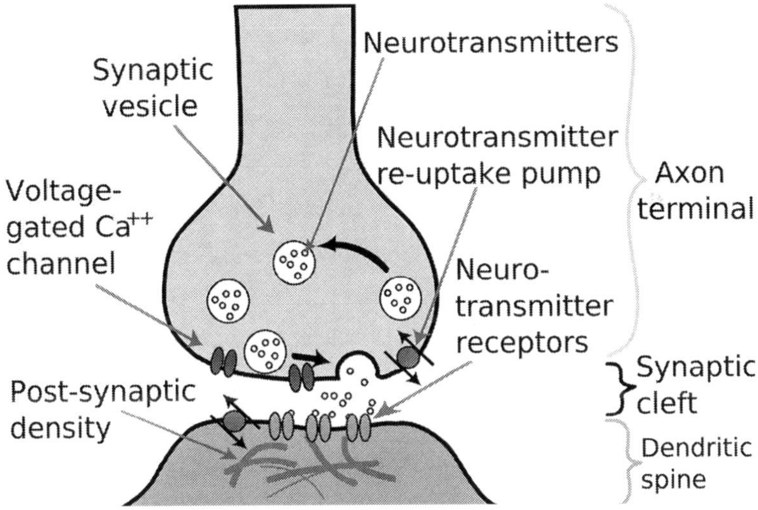

Chemical Synapse

Ironically, it is now known that this is not the case. In the 1900s scientists discovered that actually there is always a tiny gap (synaptic cleft) between neurons and this gap is big enough to stop the 'spark' jumping across from one neuron to another. This gap is called the chemical synapse **(14).** How then, does the spark get from one neuron to another? The answer is that a special and very complicated chemical process 'carries' the spark across the gap. The chemical that actually carries the 'spark' across the gap is called a neurotransmitter and the brain uses about ten different ones. Most modern medicines work at this synapse / gap by enhancing or blocking transmission here.

You might think it would be much easier to 'hard-wire' all the brain neurons together and do away with this gap; the chemical synapse. After all, this gap slows the sparks' movement between neurons and it is a complicated process. So, why not hard-wire all brain neurons together just like the electric wiring in a building? One answer is that in a house, there are switches that allow electricity to flow along some circuits and not others. When you turn on a light in your house, you don't usually want all the lights to turn on, or off, together. Similarly, you don't want all brain neurons to 'spark' together so these gaps, these chemical synapses can function like switches so parts of the brain can be turned 'on' whilst other parts can be left 'off'.

Neurons behaving like Chips

Given that the electrical 'sparks' of brain neurons travel between neurons, how does this simple 'spark' become a useful signal in the brain, especially, when all neuronal sparks are identical? One answer is to consider 'Morse code'. In the 1800s Samuel Morse an American physicist developed an electrical telegraph system which sent pulses of electricity (small voltages) along wires from a production site to a destination site. Morse used only two pulses. A short pulse a 'dot' and a longer pulse a 'dash'. He then wrote a code for each letter of the alphabet. For example the letter 'A' was dot-dash whereas the letter 'B' was dash-dot-dot-dot. The best known letters today being dot-dot-dot (S) dash-dash-dash (O) dot-dot-dot (S) that is SOS!

A •–	J •–––	S •••
B –•••	K –•–	T –
C –•–•	L •–••	U ••–
D –••	M ––	V •••–
E •	N –•	W •––
F ••–•	O –––	X –••–
G ––•	P •––•	Y –•––
H ••••	Q ––•–	Z ––••
I ••	R •–•	

Morse Code

Because each letter of the alphabet was 'coded' you can see that if the both the sender and receiver knew the code, the sender could send a stream of dots and dashes which the receiver could then 'de-code' into letters. These letters could then be grouped into words which could be grouped into sentences which could be grouped into paragraphs, pages and into a whole book. In this way, a simple code could be used to send complicated instructions from one person to another. The 'electrical' spark of the human brain neuron is just like the 'dot' of Morse code and the absence of a neuron spark is like the 'dash'. So, just like Morse code, the human brain neuron can send a stream of dots and dashes along its connections to other neurons and if both the sending and receiving neurons understand the 'code' then similarly, complicated instructions can be sent between neurons. This neuronal 'code' is not limited to neurons in the brain. All neurons throughout the whole body manufacture the very same potential or 'spark' and this code had become the language of the whole nervous system.

Computer Silicon-Chip

References and Further Reading:

11. Benjamin Franklin:
An American Life (2003)
Isaacson

12. Golgi:
A Biography of the founder of Modern Neuroscience (2010)
Mazzarello
Oxford Uni Press

13. Hodgkin and Huxley (1952).
A quantitative description of membrane current and its application to conduction and excitation in nerve . The Journal of Physiology. 117 (4): 500–544.

14. Molecular Mechanisms of neurotransmitter release (2008)
Wang
Human Press

CHAPTER 4

How Does My Brain Work?

(Brain Structure & Function)

Synopsis: This Chapter explains how the brain works and compares the size of brains in different animals. It also examples how different brain regions have different functions.

Spinal Cord

The human central nervous system comprises the brain and the spinal cord which protrudes down the back below the brain. Both the brain and spinal cord are jelly-like and fragile and will collapse under their own weight if removed from the body. The spinal cord is about the same diameter as a pencil and for protection it is encased in a bony tube. If this tube was continuous, we would not be able to bend our backs, so the tube has cuts in it which have become the vertebrae, as shown. The spinal cord has neurons which are just like brain neurons so they can also manufacture an electrical 'spark' but most of the spinal cord is a giant cable carrying the brain cell connections (axons) between brain and body, called the spinal nerves. This is much like an electrical cable in a building that has many individual wires inside it which branch out to various destinations as the cable passes around the building.

Although it is obvious that the skull protects the top and sides of the brain, the skull continues underneath the brain making it totally enclosed in a bony box. There is a hole in the bottom of this box which allows the spinal cord to protrude through and several other holes which allow various blood vessels and nerves to enter and leave the brain **(7).** Although the skull provides protection to the brain, this box also creates some problems. This is because if a tissue is damaged it commonly swells and I'm sure we have all experienced a damaged and swollen knee or ankle! Similarly, if the brain is damaged e.g. following head injury, it also swells like other tissues but because it is enclosed in a bony box, there is nowhere for it to swell into! This means that a swelling brain pushes against the internal walls of the cranium and this can damaged it, which in the extreme causes death!

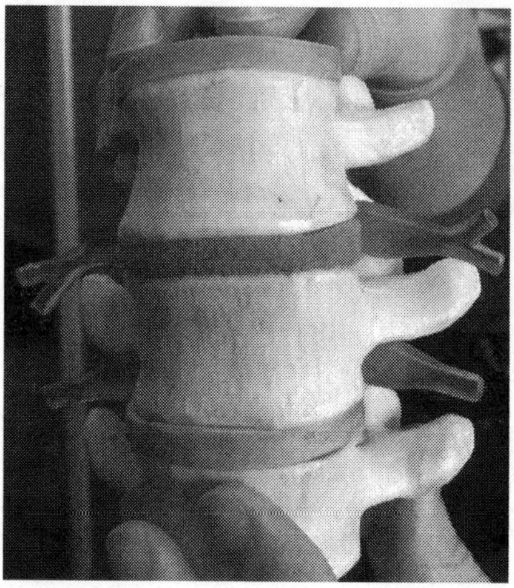

SPINAL VERTEBRAE WITH CORD INSIDE

Old & New Parts

The brain is not a single ball of millions of neurons squashed into the skull and this one ball of neurons work together on whatever task is put to it. Rather, the brain has a complicated structure with different parts undertaking different tasks. This structure has evolved over millions of years which means that our brains have older and newer parts. The most recent parts are the two large cerebral hemispheres and an old part is the Brain Stem.

Brain Stem

The brain stem is below the two hemispheres and it is an 'old' part of the brain which contains neurons responsible for maintaining life. That is, this part has specialised regions which have neurons that control the vital functions of the body like breathing, the heart beat, blood flow through the blood vessels, gut movements, our hormonal system, body fluid balance and our ability to be conscious. This core also has areas that can generate a few very basic emotions like anger. The brain stem only accounts for about ten percent of the human brain and this same region is also found in the brains of all vertebrates like horses, cows, cats and dogs. In fact, this region in humans is barley larger than the same region in a cat or dog. Perhaps this is not too surprising since cats and dogs can breathe and have beating hearts just as well as we can. In fact there are many things that many other animals can do just as well as us and often, better. For example, cats and dogs can see, hear, taste and smell better than we can. Cats can climb trees better than we can and dogs can run faster and longer than us, so the ability to do these particular skills is not what makes the human brain bigger and better.

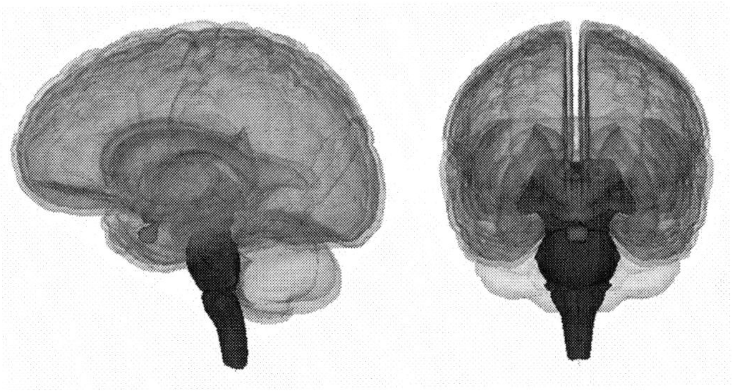

Brain Stem (Dark Area)

Who Has The Biggest One?

It is generally the case that as animals get bigger so do their brains. So, a horse has a larger brain than a cat, which has a larger brain than a mouse. This means that elephants and whales have bigger brains than us. However, their brain cell density is less than ours so although their brains are bigger, they only have about one quarter the number of neurons we have. This means that humans have more neurons in our brains than any other living creature on

Earth and furthermore, compared to our body size, our brains are larger than predicted. Presumably, our great number of neurons makes us the most intelligent creature in the entire universe! There is every reason to expect that humans and the human brain will further evolve in time and the brain will get bigger and better and I wonder what humans will be able to do in a million years from now?

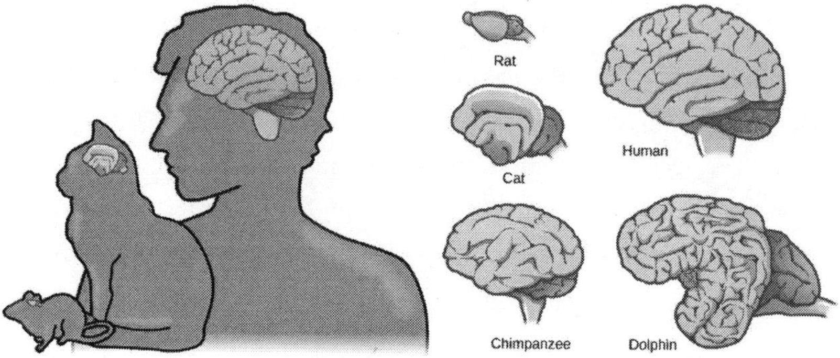

ANIMALS AND BRAIN SIZE

Cerebral Hemispheres

As human and brain evolution has progressed, the modern brain has grown two large cerebral hemispheres which are about eighty percent of the total brain and these have to be contained inside the skull. In order to occupy as little space as possible, these hemispheres have been folded, squashed, pulled and twisted and this is why the modern human brain has many folds and a wrinkled surface. If it were un-folded and un-wrinkled, our head would have to be about one meter wide to accommodate it! The reason our brain has two cerebral hemispheres is because our bodys' have evolved with 'bi-lateral symmetry'. That's is, if you drew a vertical line from your nose to the ground, you obviously see that your body has a left side and a right side and these are usually mirror images of each other. We have a left and right leg and a left and right arm etc. and we also have a left and right cerebral hemisphere. Although we commonly believe that our two feet are the same, usually they are not; there are small differences between them. There are also small differences between our two arms, hands, eyes and ears and there are also differences between our two cerebral hemispheres.

Each cerebral hemisphere has a similar structure with almost all the neurons located in the surface which forms a layer about half a centimeter thick. The connecting parts of all these neurons, the many, many axons, project down from these neurons to so that most of the mass of each hemisphere is actually axons, not neurons themselves. Even so, the number of neurons in the surface layer is still enormous, about twenty billion! Although each neuron is part of its own large network with connections to many thousands of other neurons, the neurons in one hemisphere usually connect with neurons in the same hemisphere. This is not completely true because if it was, the two hemispheres would be isolated from each other and we would have two brains rather than one. To avoid this happening, the brain has several massive 'cables' or connections that link the two hemispheres together and these ensure that each hemisphere knows what the other one is doing **(9)**. There are some people where the cables that connect the two hemispheres have been cut and this produces some unusual results, as discussed later.

It has been found that the neurons in the left hemisphere connect to the right side of the body and the neurons in the right hemisphere connect to the left side of the body. This means that all the neuron connections from the two hemispheres 'cross-over' and this happens in a region close to the back of the neck. Why we are 'wired' this way relates back to our bi-lateral symmetry and actually, there is no advantage to having each hemisphere connect to same side of the body.

Brodmann Regions

About a hundred years ago a German neuroanatomist called Brodmann studied the cerebral hemispheres and he divided the surface layer of each hemisphere, the layer of neurons, into about fifty different regions and he proposed that each of these regions performed a specific and specialised task or function **(15).** He got this idea from studying patients who had particular difficulties in life and he compared their difficulties with their medical records. For example, if a patient in life had lost the use of their left arm and their medical records showed that they had damage to a particular part of the brain, he reasoned that the damaged brain area must have been responsible for moving the left arm.

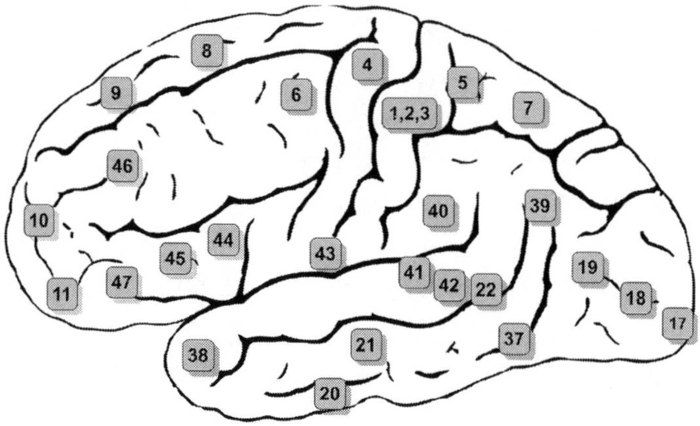

BRODMANN REGIONS

Much brain research has been undertaken since then and today science has confirmed that some of what Brodmann proposed is true. That is, some of his particular numbered regions do perform the task or function he first proposed, although many don't **(10).**

References and Further Reading:

7. Gray's Anatomy (41st Edition, 2015)
   The Anatomical Basis of Clinical Practice
   Stradring
   Elsevier

10. Principles of Neural Science (5th Edition, 2012)
    Kandel, Schwartz, Jessell, Siegelbaum & Hudspeth.
    Elsevier

15. Brodmann (1909)
    Vergleichende Lokalisationslehre der Grosshirnrinde
    J A Barth

CHAPTER 5

Let's Move It - Move It!

(Process of Movement)

Synopsis: This chapter explains how there is dedicated part of the brain which is responsible for initiating and controlling movement. Some movement difficulties are outlined.

Movement

Brodmann proposed that his region number 4 was a 'Movement' region and modern science has confirmed this to be true. 'Region 4 is also known as the Somatomotor cortex, Motor strip or Pre-central gyrus. It contains a 'map' of the body and neurons here send signals from the brain to our muscles to move our body. So for example, if we wished to kick our left leg, signals or sparks from the left leg area would send signals to make our left leg move. Remember neuron connections cross-over so this signal would come from the left leg area in the right hemisphere.

As an aside, many of these movement neurons that connect our two hemispheres to the body muscles cross-over at the same point near the back of the neck. All these crossing neurons have formed a couple of bumps on the surface of the brain and these look a little like pyramids so they are officially named the 'Pyramids'.

Further study of this movement map shows that the map is not in the same proportions as our actual body. Below is a picture of what a person would look like if they were made to the proportions of this map. You can see that the hands and lips are huge. The reason for this is that we have many movement neurons going to the muscles of the hands, mouth and throat and these require large areas of this strip. This means that we all have great control and skill over the movements of our hands and fingers, as exampled by a concert pianist! The reason we have many neurons going to our mouth and throat is that this gives us fine control over the muscles here which allows us to speak.

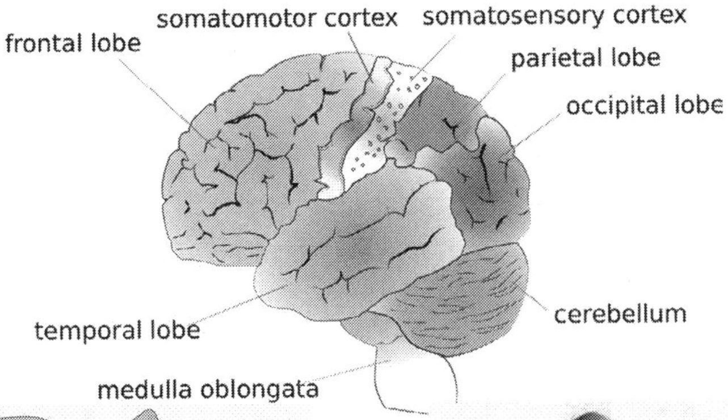

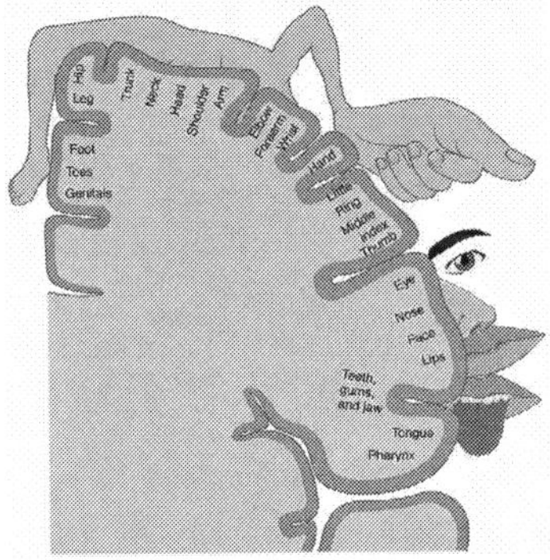

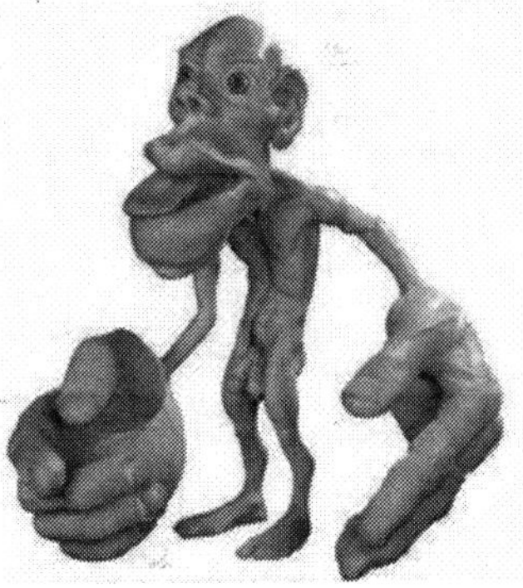

Motor Cortex

Movement Difficulties

If for some reason, this movement strip of the brain was damaged and did not work, then the signals to move our body could not be generated so our limbs could not be moved. Sadly this loss of movement or paralysis happens sometimes in people following a 'Stroke', as mentioned later. Although voluntary movements originate from this movement strip in the brain, there are many other parts of the brain that contribute to our final movements. Movement 'planning' is done in Brodmann region 5, which is right beside region 4. Movements are also monitored as they take place and this is done by part of the brain called the Cerebellum. The strength and force of limb movements are also adjusted by 'old' parts of the brain deep inside it called the Basal Ganglia. Movement then, is a complicated process and if you have a complicated and sophisticated network or system or machine and part of it fails, it might all 'crash' and totally stop working, but most likely it will continue to work but with some strange abnormalities. This is the case with movement. One example is Parkinson's disease. In this disorder the Basal Ganglia don't work properly and this causes limb tremours and body stiffness. A different failing of the Basal Ganglia causes 'Huntington's Disease' where the limbs are thrown around violently and there is muscle degeneration. Another Movement disorder is Motor Neuron Disease. In this condition many of the movement neurons in Brodmann region 4 and others in the spinal cord, slowly degenerate and most body movement is eventually lost **(16).**

References and Further Reading:

16. Neuropathology:
A Reference Text of CNS Pathology (3rd Edition, 2012)
Ellison, Love, Maria, Harding, Vinters
Mosby

CHAPTER 6

Sensational!

(Our Five Senses)

Synopsis: This chapter outlines how the brain receives our sensations of sight, hearing, taste, smell and touch. More detail is given about vision and some visual difficulties. It explains some ways the brain normally creates visual illusions.

Sensory Process

It is generally considered that we have five senses: sight, hearing, touch, taste and smell. There are arguments we have more. For example, we usually know or can 'sense' our body posture that is, whether we are standing, sitting or lying down. We also usually know or can 'sense' where are limbs are in space, that is whether an arm is by our side or out-stretched. This body-sense does not require vision, hearing, smell, taste or even touch so some neuroscientists argue this is another sense called 'proprioception'. We might have even more senses, but in all cases each sense supplies information to our brain which processes it and generates the final sensation.

Sensory Receptors

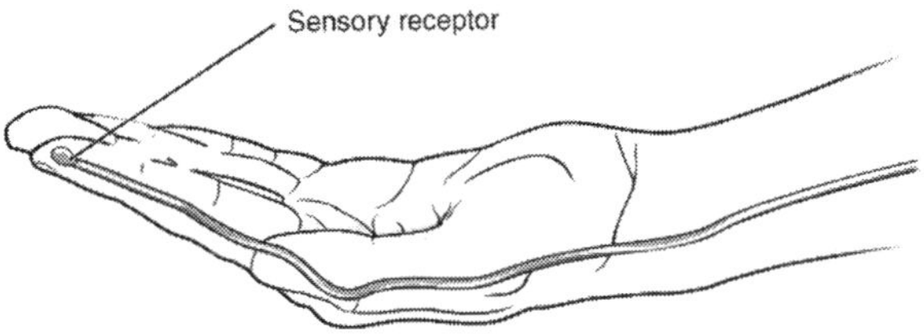

Sensory Receptor

The sensory process is much the same for all our senses. Our body has special sensory cells called receptors which can detect a particular 'sensation'. These receptors are connected to sensory neurons or nerves which connect to the brain. When a sensory receptor is triggered it stimulates the nearby sensory nerve to producing potentials or 'sparks', as described earlier. These sparks are the coded language of the nervous system and they are sent to the brain for analysis which then generates the final sensation.

All In Good Taste

Our taste receptors, known as taste buds are found in the tongue. The molecules or chemicals that make up food and drink stimulate them which makes them trigger the nearby sensory neurons to fire. These signals or sparks travel along the Trigeminal sensory nerve to the brain. They terminate in Brodmann region 43 which generates our taste sensations. Of course, if this Brodmann region was damaged, you would lose the ability to taste.

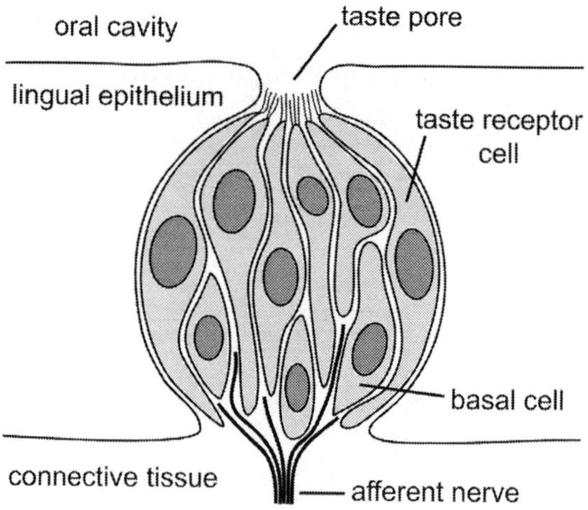

A Taste Bud

How Do You Smell?

We have smell receptors in our nose and molecules or chemicals in the air that enter our nose stimulate them which makes them trigger nearby sensory neurons to fire. These signals or sparks travel along the Olfactory sensory nerve to the brain. They terminate in Brodmann region 11 which generates our smell sensations. Of course, if this Brodmann region was damaged, you would lose the ability to smell.

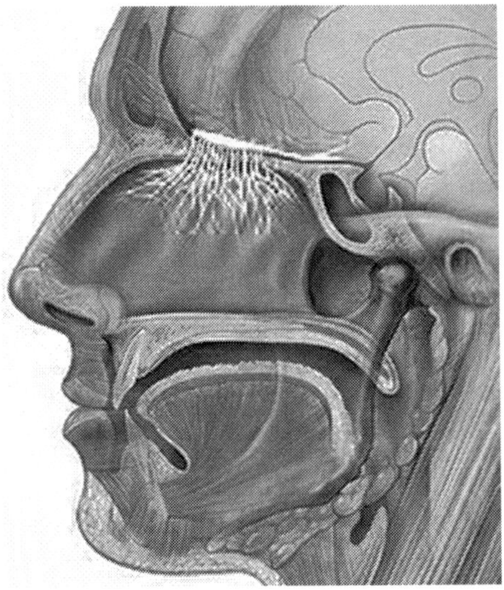

Smell Pathway

Can You Hear Me?

We have 'sound' receptors deep inside our ears and when stimulated they trigger the nearby sensory neurons to fire. These signals or sparks travel along the Auditory nerve to the brain where they terminate in Brodmann regions 41 & 42. This is rather like having a microphone which detects sounds and sends the resulting voltage to a recording device. It is perhaps no surprise that these Brodmann regions (41 & 42) in the brain are right 'underneath' the ears so that the signals generated in the ears don't have to travel very far to reach their destination site in the brain. Of course, if these Brodmann regions were damaged, we would become deaf.

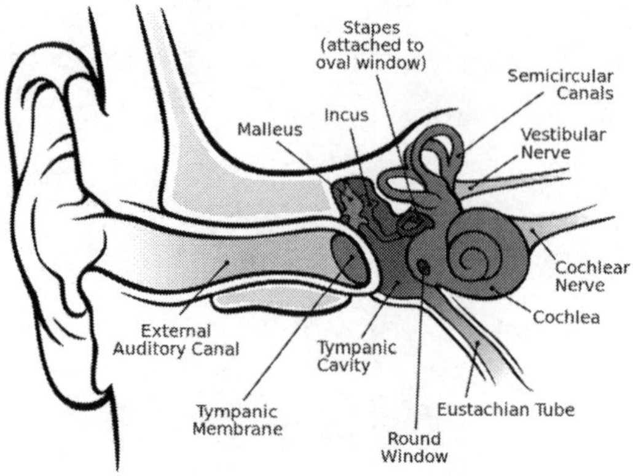

HEARING PATHWAY

I'm Touched!

Brodmann was also correct in identifying regions 1, 2 & 3 as our 'Touch' regions. These three regions are found together around the middle of a cerebral hemisphere and they are the final 'destination' site for touch sensations sent to the brain from the body. This area is known as Somatosensory cortex or Touch strip or Post-central gyrus and it has a map of the body rather like the Movement map found in region 4. This sensory system works so that if for example, you were touched on your right arm, the 'touch signals' are sent up from your right arm to the 'right arm' destination area in the brain. Remember, neuron connections cross-over so these signals would go to the arm region in the left hemisphere. When these signals or sparks arrive at this destination area we 'feel' or become aware that our right arm has been touched.

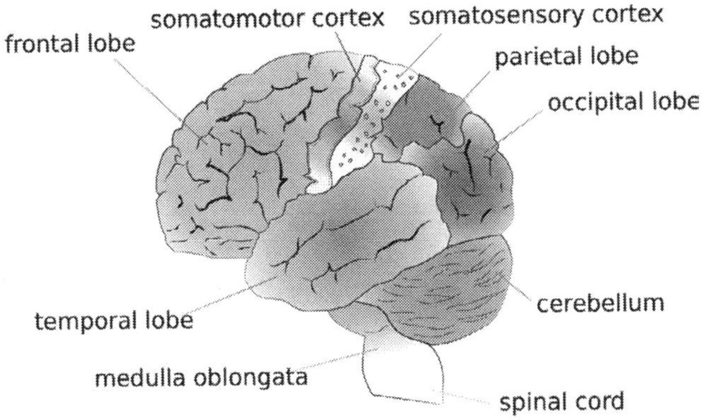

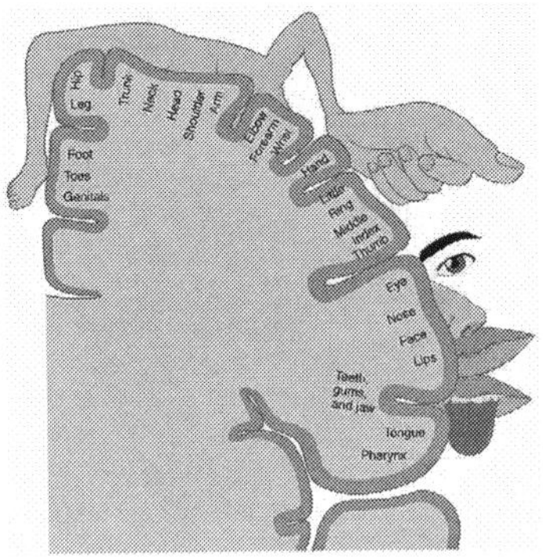

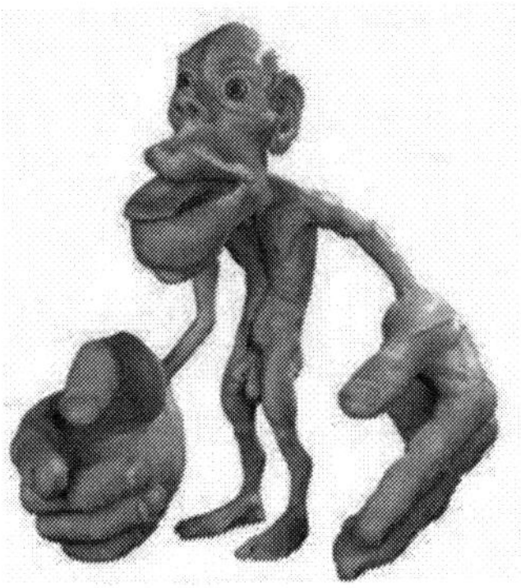

SENSORY / TOUCH CORTEX

Further study of this map shows that that it is not in the same proportions as our actual body. Above is a picture of what a person would look like if they were made to the proportions of this map. You can see that the hands and lips are huge rather like the figure for the motor cortex. The reason for this is that we have many sensory / touch neurons arriving from the hands, mouth and lips which require large areas of this sensory strip **(10).** This makes our hands and lips especially 'sensitive'.

Sensory Difficulties

If for some reason, part of this touch region did not work, we would not be able to feel anything from part of the body. Sadly, this kind of loss of sensation happens in people following a Stroke. A

Stroke is the name given to the condition when some of the blood supply to the brain has been stopped. I was lucky enough to undertake some research into Stroke at the Institute of Neurology, London. I was working for one of the Country's leading Neurosurgeons and we were trying to develop new medicines to combat the damaging effects of a Stroke **(17).** Brain neurons are not able to live for very long without a good blood supply. Other types of body cells can survive for hours without blood e.g. skin, but neurons are high-burners and they can only survive for a few minutes before they start to fail and die. Once dead, brain cells cannot repair themselves or be replaced so the function they performed is lost. It turns out, for complicated reasons, the artery that supplies blood to the sensory and motor regions of the right hemisphere, that is Brodmann regions 1, 2, 3 & 4 is the one most likely to get blocked. When this happens, blood stops flowing to the arm and hand areas in the right hemisphere and that is why most Stroke patients become un-able to feel or move their left arm and left hand **(16).**

Losing Touch!

There are a few very rare cases where people have completely lost their sense of touch following to damage to their sensory nerves that end in Brodmann regions 1, 2 & 3 in both hemispheres. In addition to all the difficulties of not being able to 'feel' anything, they also cannot tell where their body and limbs are in space which is called a loss of proprioception. These difficulties also cause them great problems with movement. Most people, most of the time can move easily and we don't really have to think about it. Sometimes we need to really concentrate on particular movements like playing a musical instrument or engaging in a sport, but usually we can walk along the street and freely talk with someone and we don't give a single thought to move our arms and legs etc. Also, we don't usually have to think about whether our movements are correct or not and this is because we have automatic monitoring systems that make any required adjustments for us.

In complete contrast, people with this massive sensory loss have to 'think' about contracting and relaxing every single muscle just the right time in order to make a movement! Even then, they can only manage to do this because they watch and monitor their movements. Visual monitoring of their movements is crucial to them because it's the only feed-back they have and if it is removed they will fall down! This whole task is commonly too much for the human brain to achieve so sadly, most people with this condition lose eighty percent of their movement abilities. However, there is a case of a man who managed to walk again and he also managed to learn how to do most normal movements. You can imagine this required enormous amounts of learning, concentration and brain power **(18).**

Rapid Messaging

You may have noticed that the 'Touch' strip (Brodmann regions 1, 2 & 3) and the 'Movement' strip (Brodmann region 4) in the brain are situated beside each other; they are neighbours. This is probably not chance. Further study has shown that there are many neurons from the touch strip that have connections direct to the motor strip next door. That is, the touch neurons of the hand area in region Brodmann 1, 2 & 3 have direct connections to the neighbouring hand movement area in Brodmann region 4, and visa-versa. This means that sensations arriving in the right-hand can send messages quickly to the right-hand movement region. You can imagine this is useful under many circumstances, such as playing the piano when the touch signals from the fingers on the piano keys need to inform the pianist where his fingers are so he can quickly execute the next movement of his fingers on the keys.

Sight

Brodmann was also correct about vision. His regions 17, 18 & 19 at the rear of the brain are the first destination sites for images coming in from the eyes. The initial process of vision follows the same pattern for the other senses. Each eye has a layer of light-sensitive receptor cells at the back and when these cells are 'bleached' by light, they cause the nearby sensory neurons also at the back of the eye in the retina, to fire. Their 'sparks' are sent along the Optic nerve from the eyes to Brodmann regions 17, 18 & 19 which start the process of 'seeing'. There are many neuroscientists who think that perhaps as much as half of the whole cerebral hemispheres work on vision.

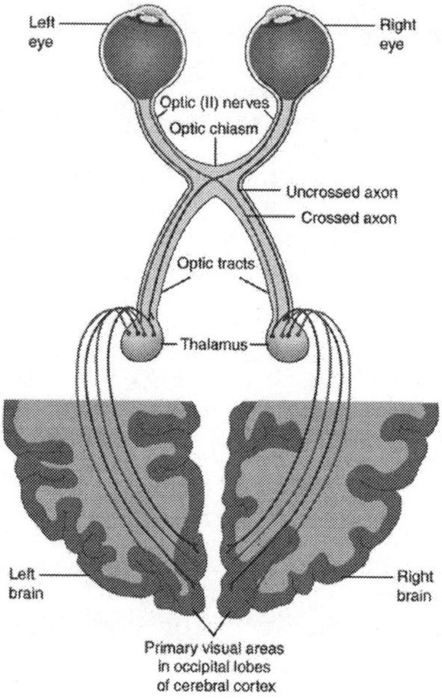

VISUAL PATHWAY

Seeing is a very complicated process which starts in the eyes and you might think that each eye sends one image to the brain but this is not so, each eye sends two. That is, four images in all from the two eyes. For complicated reasons, the image from the right-hand half of the right eye connects with the image from the right-hand half of the left eye to make one image, as shown above. At the same time, the image from the left-hand half of the right eye connects with the image from the left-hand half of the left eye. This means that the original four images sent from the eyes are eventually blended to become two images. The process is further complicated by the fact that the images sent from each eye travel to the hemisphere on the 'other' side of the brain and furthermore, all images are upside-down!

When the brain receives these images in regions 17, 18 & 19 they arrive coded as 'sparks' and they are processed. But vision doesn't stop here. It is so complicated that all the images are sent on to many other specialised visual regions found elsewhere in the hemispheres. Modern brain science believes there are at least forty of these specialised visual areas of the cortex. The early research which discovered some of these regions was done by two scientists Hubel & Wiesel and they were awarded the Nobel prize for this work in 1981 **(19).** Some of these forty regions analyse the signals for colour, others capture edges and others capture angles etc.

One major aspect of vision is to identify 'what' things are seen. Take for example the picture below of a painting by Constable in 1821 called the Hay Wain. This was painted in Suffolk, England and I'm sure you can identify objects in the painting such as the Hay Wain, the river, the dog, the man, the horse, the house, trees, fields and sky. Although we see these instantly, it actually takes an awful lot of brain processing to 'see' a shape and then we have to interrogate our visual-memory to work out what the shape is?

Constable's Hay Wain

A second major problem to solve in seeing is 'where' objects are? You can see that the trees in the distance of the painting appear rather small which could mean they are rather small trees. But our clever brains know about perspective, depth and size so they know that objects further away appear smaller compared to close objects so the brain corrects for this.

Perspective and knowledge of depth is achieved by the brain because we have two eyes in our head both pointing forwards so each receives a slightly different image of the same object. This is called bi-nocular vision. These two slightly different images are superimposed by the brain which is able to construct 'depth' and where things are. You can get an idea of a two-dimensional world by looking at objects with one of your eyes covered. You still get some idea of depth, but much more if you un-cover the other eye. Two eyes are certainly better than one. This is why most animals have two eyes positioned in the front of the face. Most birds don't have this, their eyes are on the sides of their head so they can't get much visual perspective when they see. However, birds of Prey need to see their prey accurately so they have both their eyes facing forwards. Two eyes are better than one and three eyes would be better than two, but we have not evolved that far yet **(10).**

These forty brain regions that analyse the visual signals from the two eyes report back to a final brain region that collects all this information and constructs the final complete image that we 'see'. It must be remembered that we live in a moving world so we are not talking about our brains being able to simply construct a few 'still' pictures over a few seconds. Rather, we have to make a complete movie in a fraction of a second, repeatedly! Quite astonishing; and science has not yet worked out how the brain manages to do this so quickly!

Colour Vision

Although most of us see in colour, not everyone does. There are various reasons why some people can't see in colour and seeing normally in colour is not as simple as you might think. Firstly, our eyes can't actually 'see' different colours, but they have methods to detect lights of different wave-length which they en-code and send signals to the brain which is able to de-code and 'artificially' construct the colours that we 'see'. After this, our brains do some interesting comparisons so that when we 'see' a red cup, we don't just see the cup as red; because it is red. Our visual processes compare the colour of the cup with the colours of other surrounding objects. Then our brain makes a comparison between the red cup and the other surrounding objects and then it finally decides that we see the cup as red. This means that if you saw the same cup with different surrounding colours, you might not see the cup as the same red!

If you think this unlikely, take a good look at the central horizontal band / line in the picture below and you will see that the left-hand side of this line is dark grey and this gets lighter grey as you move along it to the right.

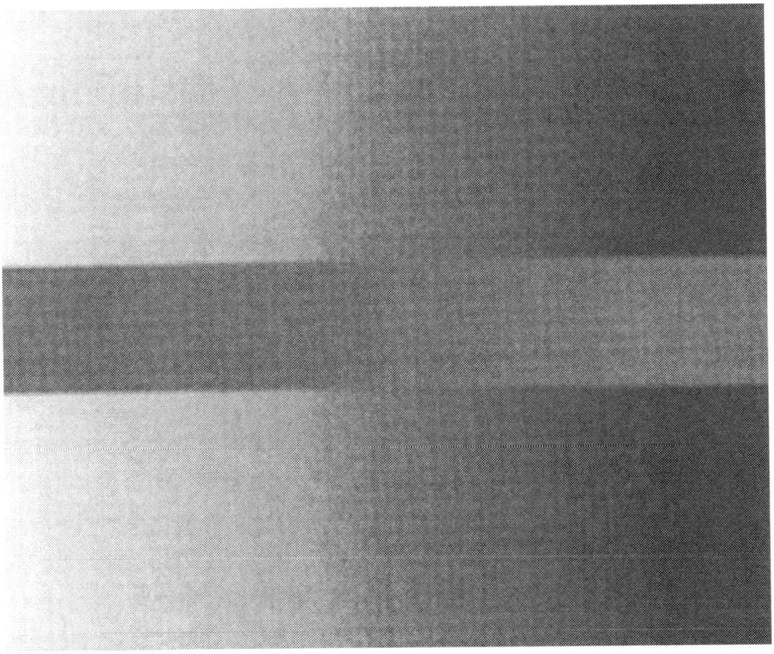

Grey-Scale Illusion

Actually, the colour of this grey line doesn't change at all. To see this, cover all this picture except the central grey line / band and you will see it is a constant grey colour all along its length. The reason we see it change is because, as mentioned, our visual processes are comparing it with the surrounding greys and this creates the visual illusion of colour change! A video camera would not be 'fooled' by this and it would not see the 'illusion'.

Another example of clever visual programming by our brains relates to seeing human faces. Neuroscientists think that one of our forty special visual regions is designed and dedicated to seeing human faces. This region exists because we see so many human faces in life which are all rather similar; that is, all faces have two eyes, a nose and a mouth and we might find it very difficult to tell them apart. All people might, therefore, look the same. So, to allow us to distinguish between them, we have this additional and special face-recognition region. This region doesn't work on 'upside-down' faces **(10)** so if you look at the face below and you might take a while to recognise who this is?

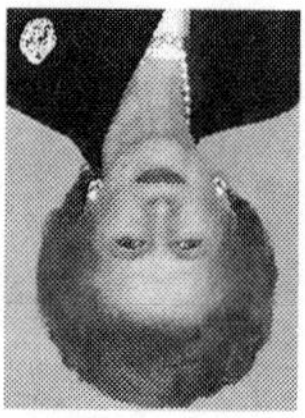

Famous Lady

But you will be able to recognise this man instantly; because his face is the right way up.

Famous Man

Seeing Is Believing?

This special face-recognition region can cause other effects because it is constantly trying to see faces in the world around us. Sometimes it will convince us that we see a face when it is not really there. For example, put this book down and look at the picture below from five meters away.

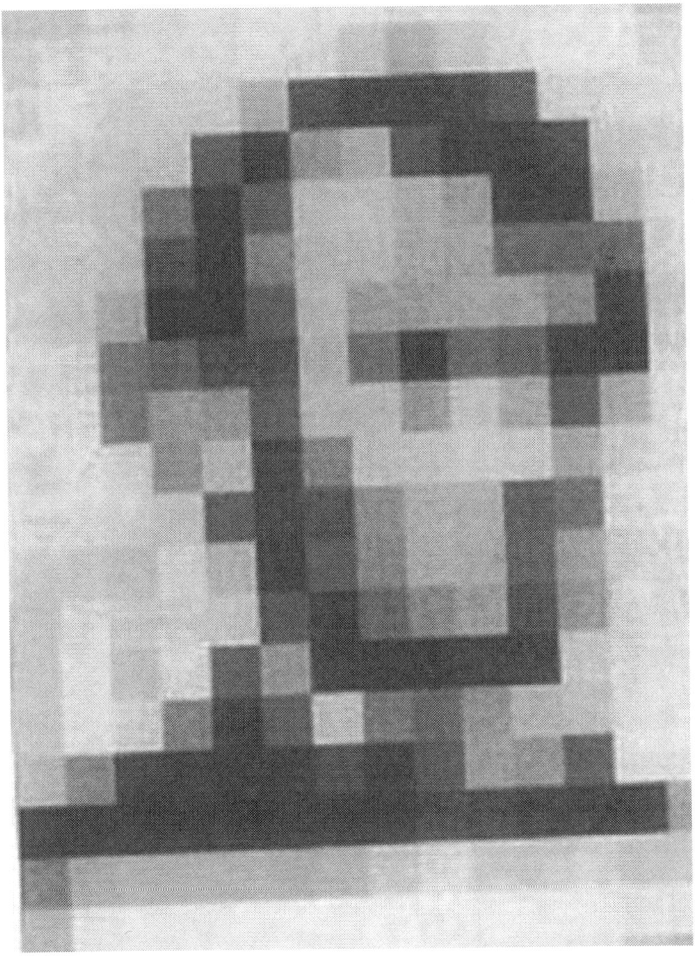

WHO IS THIS?

I bet you see a picture of Abraham Lincoln who was President of America around the 1850s. But if you get to the normal reading distance from this book, all you will see are pixelated squares and not a face at all, let alone Abraham Lincoln. A camera would never see Abraham Lincoln because it is not able to create images like our complicated brains can. This makes you wonder how much of what we see, hear, smell, taste or touch in the outside world is really there and how much is made-up by our own brains?

Synaesthesia

I have mentioned that each of our five senses; taste, sight, hearing, smell and touch has its own 'destination' site in the brain which produces the final sensation or perception. Touch for example, going to Brodmann regions 1, 2 & 3. These five senses are normally kept separate and do not 'overlap' although some people claim that some of their senses do merge. For example, the arrival of sounds in these people will also stimulate vision so the person not only hears the sound, they also 'see' it as swirling colours. Similarly, the arrival of a particular image can stimulate the person to 'smell' it. These unusual sensory connections are called Synaesthesia. It is difficult for neuroscientists to confirm that these strange sensory connections actually happen but in support of it, many psycho-active drugs appear to be able to make similar, unusual sensory connections, as sung by the Beatle the late John Lennon.

Picture yourself in a boat on a river
With tangerine trees and marmalade skies
Somebody calls you, you answer quite slowly
A girl with kaleidoscope eyes

Cellophane flowers of yellow and green
Towering over your head
Look for the girl with the sun in her eyes
And she's gone

Visual Difficulties

Like other senses, vision results from a complicated and sophisticated network or system and if part of this fails, it might all 'crash' and totally stop working, but most likely it will continue to work but with some strange abnormalities. This is the case with vision and you can imagine that if some of these forty regions did not work properly, our vision & seeing would become abnormal. Sometimes following head-injury or a Stroke, part of the visual system is damaged with unfortunate effects. Total blindness would result if Brodmann regions 17, 18 & 19 stopped working and there are many cases reported of patients who have damage to some of the other visual areas.

What's That?

I have mentioned that one major aspect of vision is our brains being able to work out 'what' things are that we see. This relies on us interrogating our visual memory and matching what we see with our knowledge of the visual world. There are a few cases of people who have damage to their 'what' system. They can see the world outside perfectly well and describe things in detail, but they can't work out what they are looking at! For example, one person who looks at a country scene of fields, trees and sky describes what they see as 'A lot of green at the bottom with some green shapes pointing up into the blue region up above'. The green below being the grass, the green shapes pointing up are trees and the blue region above is the sky. If you ask them 'What is a tree?' they can tell you well enough and describe it but they can't recognise a tree when they see one! It often helps if they can feel an object to get clues about what it is.

Face-Blindness

The brain region thought to be our special 'facial-recognition area' is called the fusiform gyrus and damage to this area causes a condition called face-blindness or Prosopagnosia. People who have this condition can look at a face and see the eyes and look again and see the mouth and look again and see the nose etc. but they can't put all these images together and see a whole face. Such a person can't recognise any one by looking at their face because they can't see the whole face and in the extreme, these people can't recognise a picture of their own face! Sadly, these people can't live normal lives. For example, if they want to meet a friend somewhere, they have to agree a place and time, as we all do, but because they won't be able to recognise the face of their friend, they have to agree beforehand that the friend will be wearing certain coloured clothes or wearing a particular kind of hat!

Only Seeing Half The World

Neglect is a condition that sometimes happens after a Stroke. It is a condition when the person only sees one side of the world in front of them, say the right-hand side and the left-hand visual world is missing! This is not a problem with the eyes but with the brain.

NEGLECT DRAWINGS

When a person with Neglect is asked to copy a drawing as above, they only see one side of it (the right-side) so they draw everything on the right side 'neglecting' the left side. In life, they always see the visual world on their right-hand side and nothing on the left. To correct this and place things on the left, they turn to the right, but this doesn't work, so they turn more and more to the right and they can end up walking around in circles! **(16).**

Stationary Or Moving?

Akinetopsia is an extremely rare condition where patients have profound motion blindness following brain injury. They start to see the world as a series of still photographs that change, but they can't link them together to see life as a 'movie'. Life is very difficult and you can imagine they can't deal with anything that moves! Cars jump along the road and people jump around them! In contrast, we can sometimes see things move when really they are stationary. Look below, and watch the cogs turn ....

Turning Cogs

References and Further Reading:

10. Principles of Neural Science (5th Edition, 2012)
Kandel, Schwartz, Jessell, Siegelbaum & Hudspeth.
Elsevier

16. Neuropathology: A Reference Test of CNS Pathology (3rd Edition, 2012)
Ellison, Love, Maria, Harding, Vinters
Mosby

17. Symon, Branston, Chikovani and Pullen. (1979). Ischaemic brain oedema following middle cerebral artery occulusion: Relationship between regional water content and blood flow at 1 to 2 hr. Stroke 10, 184-191.

18. Losing Touch: A man without his body (2016)
Cole
OUP Oxford

19. Eye, Brain and Vision (1995)
Hubel
Scientific American

CHAPTER 7

Ouch, That Hurts!
(Pain)

Synopsis: This chapter explains how and why we have pain and how it is received in the brain. The subjects of pain-killers, anaesthetics and pain difficulties are also covered.

Pain

Pain is an unpleasant experience and there is the question of 'Why do we feel pain?' The reason is that pain is a signal that informs us of damage to our body and it warns us not to repeat the experience. Although you might think it would be rather good not to feel pain, this has risks. There are people who don't feel pain in various parts of their body and this can cause great difficulties because they can have serious, even life-threatening damage to their 'numb' bodies and not know it, or not care about it! Of course, there are occasions when its best to remove pain, like having your teeth pulled, but generally pain is experienced by all animals and it has evolved as natural signal which is protective and necessary **(1).**

Neuroscientists are divided as to whether pain is a separate 'sense' in addition to the standard five of touch, hearing, taste, smell and sight, or not. One reason for this debate is that pain can be experienced from extreme stimulation of any of our accepted five senses. For example, very loud sounds can be painful and so can intense light, intense smells and extreme tastes. This suggests pain is not a separate sense but it is the intense experience of existing ones. However, further study has shown that our body has separate 'pain' nerves and these are entirely separate from our other 'sensory' nerves. Also, these pain nerves follow different pathways and terminate in different regions in the brain.

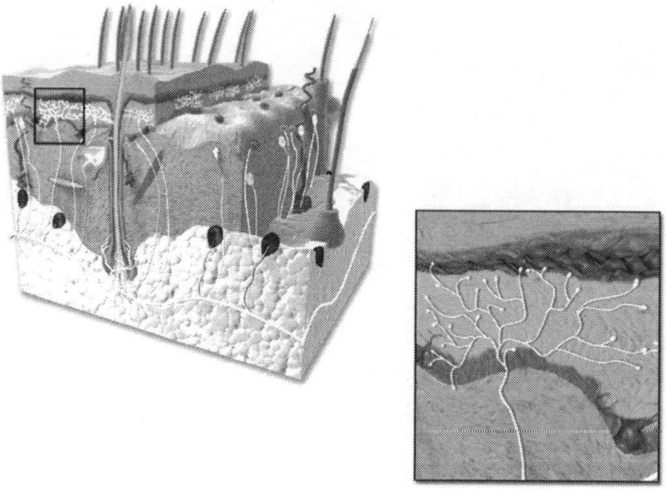

FREE NERVE ENDINGS

Take skin as an example. There are various different kinds of 'touch' receptors in the skin eg. Pacinian corpuscles. Some of these receptors are sensitive to vibration others to stretch and others to heat etc. These receptors link to sensory nerves that convert the sensation into neuron sparks that are sent to the touch region in our brain in Brodmann regions 1, 2 and 3. It is here that the signals are received and we 'feel' our skin has been touched. These sensations, therefore, use the same common sensory process mentioned earlier. But, in the skin there are separate pain / free nerve endings and when these are stimulated they send their own neuron sparks to the brain where 'pain' is felt, see image. These pain nerves are not just found in the skin, they are present over most of the body, inside and out, which allows us to feel pain virtually everywhere in our body. One notable exception to this rule is the brain which has no pain nerves so we can't feel pain in the brain itself. We can feel pain in our face, scalp, skull and head; that's why we can have a 'headache'; but there are no pain nerves in the mass of the brain neurons themselves. This is why neurosurgeons can operate on the brain without the need to anaesthetise it.

Pain Pathway

I have mentioned that we have a common sensory process in our body whereby sensations are first detected by sensory receptors which trigger nerves to produce 'sparks' which code the sensation and these are sent to the brain for de-coding and processing. Before these signals reach the brain these sensory nerves enter the spinal cord which is rather like a giant cable containing thousands of wires travelling to and from the brain. Although I have portrayed the spinal cord as a massive cable and this is generally true, the spinal cord does get significantly involved with processing pain signals before they reach the brain. In the 1960s neuroscientists discovered that some neurons in the spinal cord operate like a pain 'gate' **(20).** When the gate is open, pain signals are allowed through and they travel on to the brain where they are received and felt. Alternatively, these spinal neurons can close the gate so that pain signals from the body are stopped and they cannot travel up to the brain so pain is not felt.

Usually, this spinal pain-gate is open so we feel pain but there are situations that can close it. A simple example is 'rubbing the surface of the skin' and this probably why after we have received a blow to the skin which is painful, we commonly rub the skin with our hand which reduces the pain. Some anaesthtics can close this spinal gate and this is how an 'epidural' anaesthetic works during child birth. The anaesthetic is introduced close to the spinal cord and from here it diffuses into the spinal cord where it closes the spinal pain-gate which stops the pain signals reaching the brain.

Pain Destination Site In The Brain

In the case of our five main senses, it is established that these sensory signals are sent to specific destination sites in the brain. An example being touch signals are sent to Brodmann regions 1, 2 & 3. However, there does not appear to be a 'pain' destination site in the brain which has a pain map of the body so neuroscientists remain uncertain where our pain signals go? A few brain areas are candidates which includes the Insula cortex and peri-ventricular areas but although pain must be processed somewhere, it is not clear where! There is another important aspect about pain and that is, it has an enormous psychological component. When we feel pain in addition to it being an unpleasant physical experience, it also has great psychological importance. Pain usually worries us and we commonly consider what are the consequences of this pain and the associated damage to our body? In the extreme, will we live!

Pain Killers

We have a 'pain pathway' which starts at the pain nerve endings in the body, e.g. the skin and it continues to the spinal cord and on to the brain where we finally feel pain. Analgesics or Pain Killers are drugs that can block this pain pathway in a number of ways and in a number of places. Pain Killers can work on the pain nerve endings at the start of the pathway and this stops the signals reaching the spinal cord. A common drug that does this is Paracetamol. Another example is the 'anaesthetic' used by dentists which is injected into the gums. 'Spinal' anesthetics are injected near the spinal cord and these shut the pain-gate half-way along the pathway in the spinal cord as exampled by an epidural in child-birth. Central anaesthetics work at the end of the pathway in the brain itself. Some of these pain killers don't actually stop pain, but they put the brain or person into a 'state-of-mind' in which they feel pain, but they just don't care about it. Morphine, Opium and Heroin probably work this way. Finally, there are general anaesthetics which work by shutting down the whole brain and removing consciousness which of course, stops us feeling pain.

Natural Pain Killers

It has been known for several thousand years that opium can remove pain and in the 1970s neuroscientists discovered that the brain produces its own pain-killers which are just like opium. These substances are called Endorphins and they are manufactured in the brain and are released under various circumstances. It is likely that the release of endorphins explains how some people are able tolerate acts like fire-walking and hanging from hooks!

PAINFUL PASS-TIMES!

Pain Difficulties

Most of the human body has pain nerves. The brain tissue itself does not have pain nerves and some other tissues and organs have unusual pain 'wiring'. The heart is one example. The pain nerves in the heart cross with some of those for the left arm which is why when someone has a heart attack, which triggers pain nerves in the heart, the pain is commonly felt in the left arm! This is called 'Referred' pain. Other examples of referred pain are the Pancreas where pain in this tissue is commonly referred to the back and pain in a lung is felt in the left shoulder.

A rather odd pain difficulty is called 'Phantom Limb' pain. I have mentioned that if you damage a complicated and sophisticated network or system, it might all 'crash' and totally stop working, but most likely it will continue working but with some strange abnormalities. This is the case with pain. When a person has had a limb amputated, they might feel pain in the absent 'phantom' limb! It seems that our brain expects to receive sensory signals from all our limbs all the time so if a limb is lost, this disturbs the brains sensory expectations and the brain creates pain in the missing limb. This pain can be disabling and often anaesthetics don't work making the pain impossible to stop!

MIRROR METHOD

One method which sometimes stops the pain is called the 'mirror method'. In this method, a mirror is placed near the body which reflects an image of the other, present limb back to the person in the position of the missing limb. When the person sees the reflection of the present limb where the missing limb should be, it often convinces the brain that the missing limb has returned, so the pain stops!

References and Further Reading:

1. Berne & Levy. Physiology (6th Edition, 2010)
   Koeppen & Stanton
   Mosby
   Elsevier

20. Pain Mechanisms: A New Theory (1965)
   Melzak & Wall
   Science, 150, 971-979

CHAPTER 8

Can We Talk About It?

(Language)

Synopsis: This chapter covers how the brain gives us the ability of language and memory. Some language difficulties are discussed.

Language Areas

Modern brain science has since discovered that there are more Brodmann regions in the brain that perform the task Brodmann first thought about a hundred years ago. These are Broca's Area and Wernicke's Area, both of which are crucial for language **(10).**

Wernicke's Area – What Did You Say?

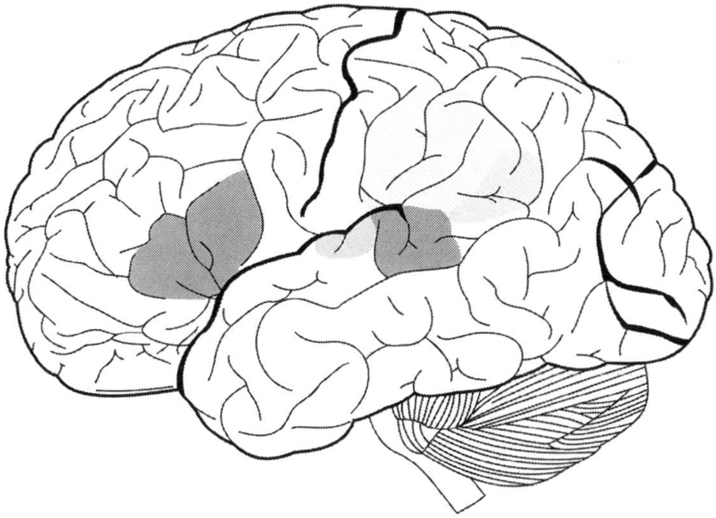

Broca's Area (Dark area on left)
Wernicke's Area (Dark area on right)

Karl Wernicke was a German anatomist who lived around 1900. He studied patients that had brain damage and who could no longer understand their natural language. He discovered this was due to damage in the cortex in Brodmann regions 20, 21 & 22 but only in the in the left hemisphere. Neuroscience has confirmed that this region is crucial for us to learn and understand a language. As we grow and develop, we naturally hear language spoken around us and our brain locks into it. At first, spoken words are just sounds or meaningless noises but we start to realise these noises mean something and we learn to speak. This ability is due to the working of Wernicke's region and without it everything would sound Double-Dutch! It's not too surprising that this region is next to our main auditory destination site in Brodmann regions (41 & 42) so the sound signals received from our ears can be rapidly sent on for analysis. Although these patients can't understand what is being said to them, they often retain the ability 'speak' so this damage causes a listening difficulty rather than a speaking one!

Broca's Area – I'm Speechless!

Pierre Paul Broca was a French physician who lived around the same time as Karl Wernicke. He also worked with patients who had language and speech problems following brain damage and he discovered that his patients had damage in a particular region of the left frontal lobe. His patients could understand what was said to them, but they could not formulate words properly in their 'minds' and they could not speak them. Neuroscience has confirmed this area is responsible for allowing us to transform our 'wish to speak' into actually being able to move the mouth and throat and lips to say words and this brain area corresponds to Brodmann regions 44 & 45.

Speak With Many Tongues!

Broca's and Wernicke's areas together provide us with a rich spoken language or languages if you are multi-lingual and this ability is unique to humans **(10).** Although many other animals communicate with each other; dogs can bark and cats can meow, no other animal has our rich spoken language and vocabulary. Even our close relative the chimpanzee, can only achieve a vocabulary of a one to two year old child. Our rich language has developed because of two important evolutionary advances. Firstly, to be able to make the required subtle sounds of words, evolution has changed the shape of our throat, voice-box, vocal cords, mouth, checks, tongue and lips and we have fine control of our breathing. All these put together means we have the anatomy required to speak compared to a monkey which can only grunt and scream.

The second important evolutionary change was the enlargement of the frontal lobes and the inclusion of Broca's and Wernicke's regions which interestingly, reside only in the left hemisphere making the right hemisphere nearly mute! These two regions are virtually absent in monkeys and our distant ancestors such as Homo Erectus **(9).** These brain regions have given us the brain power to be able to formulate words in our minds and understand grammar, syntax and semantics. That is not to say, that we all intellectually understand what nouns, consonants, verb-conjugation and past-participles are; children certainly don't; but we all have the brain power to naturally and spontaneously develop language as children do in early life thanks to these two brain regions.

Reading

Reading is usually learned along side speech at an early age. Our brains are also pretty clever when it comes to reading since most animals can't achieve this. Some people have reading difficulties and the most well-known one is Dyslexia. It is commonly thought that Dyslexics read letters backwards so they have trouble distinguishing between 'p' and 'q' or 'b' and 'd' and although this does happen, the condition is broader than this. Dyslexics have trouble with general spelling, reading quickly, writing words, 'sounding' words in their minds and pronouncing words when reading aloud. They might also have trouble reading this, but you should be able to…….

It is spuirisng that we can raed wrdos qtiue wlel so lnog as the fsrit and lsat lrettes of ecah word are ccerrot. Our cevelr biarn can qlkciuy wrok out waht the wdros rllaey are. Semitemos our bairn can eevn do tihs wtih two ltteer wdros lkie si, ro, and ti, when the lrettes are resrevd!

Our language acquisition is a massive evolutionary gain because it allows us to understand and be understood and we can pass on our knowledge which can be recorded. Indeed, many philosophers think that it is our complex spoken language that makes us unique, civilised and human.

Language Difficulties

As mentioned, if there is damage to a complicated and sophisticated network or system it might all 'crash' and totally stop working, but most likely it will continue working but with some strange abnormalities. This is the case with language. Damage to Wernicke's or Broca's areas causes significant difficulties and there are other brain regions that contribute to language and damage to these regions also produces difficulties known as Aphasias.

Aphasias

Individuals with receptive aphasia may speak in long sentences that have no meaning. They may add unnecessary words and even create new ones. For example, someone may say 'You know that winkerdon says he was snoodled outside but cold perhaps?. They actually meant 'The dog needs to go out for a walk but it is too cold perhaps? Individuals with anomic aphasia have difficulty naming certain words linked by their grammatical type e.g. difficulty saying nouns and naming things. For example, if you held a pen up in front of a person with anomic aphasia and asked them 'Can you name this? They would reply 'its a thing you write with' If you replied 'Yes, you are right, you do write with it, but what is it called '? They would reply 'You make a letter with it'. They just could not 'name' it. The same person would know everybody in their family but they could not say the name of any of them! This aphasia is commonly seen in people with dementia. People with Expressive aphasia speak short phrases that take great effort. They commonly omit words like is and it and the. For example, a person with expressive aphasia may say 'Good you' when the mean to say 'It is good to see you'.

Other Brodmann Regions

When all the Brodmann regions known to have functions were added together by scientists in the 1950s, they realised that together they only accounted for about twenty percent of the total surface area of the two cerebral hemispheres. This meant that they were not sure what the other eighty percent did? This led to the proposal that we only use twenty percent of our brains and although modern brain science has shown this not to be the case, there are still people to-day who continue to make this wrong statement.

Memory

I have mentioned that all are sensations and experiences that enter the body are coded into electronic 'sparks' which are the language of the nervous system. This code is much like the code that computers use called machine code, see later. In machine code, the computer-chip manufactures a sequence of small electrical voltages and this sequence can be stored in the computer's memory and later re-called. Our human memory is thought to work in a similar way. When we see an image, our eyes create a series or array of sparks which codes the image and these are sent to the brain. The brain is able to record this pattern or sequence of sparks and store it in our memory which can be re-called later. Although computer scientists know how computer memory stores the pattern of voltages, neuroscientists do not know how the brain stores the pattern of neuronal firing! We do know that some parts of the brain are important in this process and one structure is the hippocampus. This structure is deep in the brain and it is shaped like a 'seahorse' (Hippocampus is Greek for seahorse) **(7).** Somehow, the hippocampus along with other brain regions are able to store these patterns and produce memories.

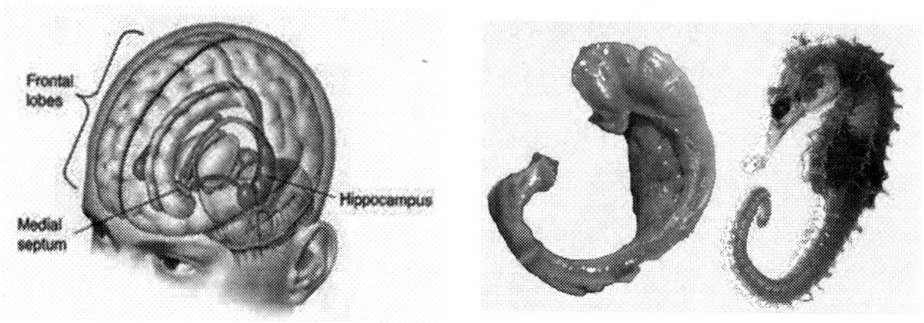

HIPPOCAMPUS

Memory Difficulties

There are several different kinds of memory loss or amnesias. For example, some people can't remember recent events called 'short-term' loss and others can't remember distance events called 'long-term' memory loss. Some people can't remember to do things in the future whilst others can't create new memories. For some people amnesia can ruin lives as exampled by not being able to remember what you spouse looks like! In this situation every time the person meets their spouse it's like meeting a stranger for the first time, again and again! Deterioration of memory is common in the elderly and it often a symptom of dementia due to degeneration of the hippocampus. A person's short-term memory usually fails first but as the disease progresses their long-term memory is also lost. Dementia patients can lose the memory of their whole lives and the people in it and there is no treatment for dementia at present.

References and Further Reading:

7. Gray's Anatomy (41st Edition, 2015)
The Anatomical Basis of Clinical Practice
Stradring
Elsevier

9. The Ancestor's Tale (2016)
Dawkins & Wong
Clays

10. Principles of Neural Science (5th Edition, 2012)
Kandel, Schwartz, Jessell, Siegelbaum & Hudspeth.
Elsevier

CHAPTER 9

One Or Two Brains?
(Left and Right Hemispheres)

Synopsis: This Chapter explains how the brain has a left and right cerebral hemisphere. It discusses how these two hemispheres have slightly different functions and it outlines some established sex differences in the two hemispheres. The possibility of having two brains is discussed.

Cross-over

Our bodies are bi-laterally symmetrical and we have two cerebral hemispheres where the left hemisphere connects to the right side of the body and the right hemisphere connects to the left. Our cerebral hemispheres are mirror-images of each other so that each performs the same tasks for the other side of the body. However, brain science has discovered that the two hemispheres are not exactly equal. There are some tasks and functions that are performed only one hemisphere and not the other.

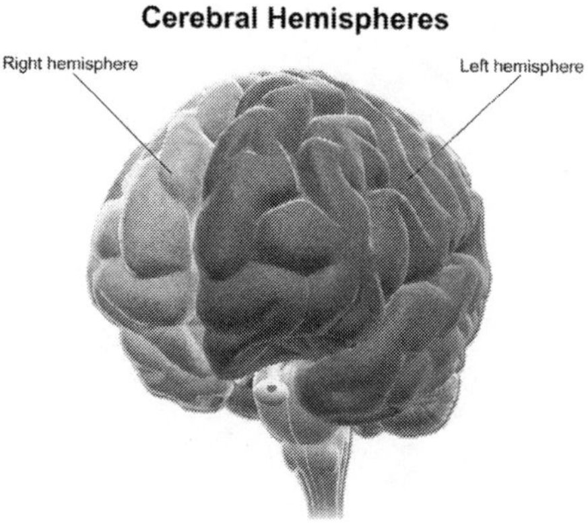

Left And Right Brain

Language

I have mentioned that the brain regions that greatly contribute to language namely, Broca's and Wernicke's are only found in the left hemisphere making the right hemisphere almost 'mute'. This is certainly true for men although in women these regions are also found in the right hemisphere. Why these functions should be restricted to only one hemisphere, is mentioned later.

Give Him A Hand!

Perhaps the best know example of hemispheric differences or 'dominance' is 'handedness'. About ninety percent of us are right-handed which means that we can perform movement tasks better with our right-hand than our left hand. This means that the right-hand movement area which is in the left hemisphere, is better than or dominates the equivalent left-hand area in the right hemisphere. There are various ways to test handedness. Some of the simplest tests consider which hand you write with or hold a knife with to eat; or hold a tennis racket or golf club. If we are right-handed we are also likely to be right-armed and right-legged. That is, you will probably throw a ball with your right arm and kick a ball with your right foot. We are also likely to be right-eyed; that is, we prefer to look through a telescope with our right eye.

This handedness affects other aspects of our lives since we are also likely to be right-lots-of-things, like when getting dressed a right-handed person will usually put their right shoe on first, put their right leg into a pair of trousers first, put their right arm into a jumper first and put their right sock on first. They will also fold their arms so that the right arm is underneath their left arm as it crosses the chest. If you don't believe me; try for yourself. Also, try to do these tasks the other way around. That is, try to put your left arm first into a jacket or left leg into trousers first when getting dressed. Fold your arms the other way and you will find all these rather difficult!

If you try some of these simple tests, you are likely to find that you are not 'purely' right or left handed. Most of us are a mixture with one hand dominating many tasks, but not all. Some people might write with a pen in their right hand but hold a knife with their left hand to eat a meal. A small minority of people are 'ambidextrous' that is they can use both hands equally well and this is clearly an advantage.

Most animals like cats, dogs, horses and sheep etc are not handed, only humans and some monkeys are handed and brain scientists are not certain why this is so. One reason might be that we don't have enough room in our brain to allow us to be ambidextrous so our brains have decided that remaining single-handed can free-up some brain space to allow us to have other useful skills. There is evidence to support this theory as mentioned since Broca's and Wernicke's regions are located only in the left hemisphere, making the right hemisphere nearly 'dumb & 'mute'. This has freed the right-hemisphere to deal with emotional intelligence, empathy and creativity **(10).**

Men versus Women

There appears to be sex differences in our brains. There are clearly many differences between men and women, but having different brains is perhaps not so obvious. The very presence of male and female versions of creatures like us probably evolved millions of years ago with the arrival of the 'sex chromosomes'. In us, there is one pair of chromosomes the 'sex chromosomes' that determines our sex, man or woman. This is often known as the XY system. Genetically, at one time very early on all humans-type creatures were 'women' who had two X chromosomes so they were XX. At some point a long time ago, one X chromosome 'mutated' into a smaller Y chromosome and having this Y chromosome and being XY, resulted in the mutant male **(9).**

Modern genetics has established that this male chromosome, the Y chromosome, is responsible for the production of a hormone called Testosterone and this hormone causes the female embryo to become male. Men have many obvious physical differences compared to women such as they are commonly bigger, taller, heavier, more muscular and stronger. It is likely that testosterone also shapes the male and female brain to be slightly different. It is not so easy to see these differences by simply looking at a male or female brain but the effects of these differences can be seen in some skills and abilities.

For example, most men compared to most women have better eye-hand skills, logical thinking and spacial awareness. For example, most men are better than most women at playing snooker, throwing darts and driving cars. Of course there are many women snooker players who can easily beat many men, but the World's best are men. Similarly, there are many women who can drive cars better than many men, but there are not any women f1 drivers on the start grid. In terms of logical thinking, there are about 1,500 Grand Masters at chess in the world to-day, but only thirty are women. These skills do not require great physical strength so this alone cannot explain the male dominance **(21).**

In contrast, most women have superior emotional intelligence compared to men. That is not to say that no man can understand how other people are feeling, but women usually understand far better. Also, women have superior negotiating skills than men. I am not saying that no man can negotiate with anyone, but history shows that male negotiation usually ends up as a physical contest and often with a war. Women just do it better. Women also have a greater vocabulary and are better readers. Vive la difference!

Having Two Brains

The two cerebral hemispheres are connected by several large 'cables' and signals along these allows the two hemispheres to 'talk' to each other. There are some extreme medical conditions like epilepsy (see later) that can only be treated by cutting the connections between the two hemispheres! This has medical benefits but it also usually produces some strange effects for the person.

People who have had their connecting cables cut have a 'split-brain', or two brains, so that each hemisphere receives its own sensations, has its' own thoughts and makes its' own decisions. You might think this situation would make a person have the experience of being two people, but this seems not to be the case. Most of the time they live normal lives because one or other hemisphere dominates so they can move and feel relatively normally. However, there are some circumstances when things are not normal. If such a person is shown an image in their left visual field only, the image goes only to the right hemisphere (remember neurons cross-over) and the person is able to 'see' the image. But because the right hemisphere is 'mute' and the image is not passed across to the 'speaking' left hemisphere, the person cannot say what they have seen! Similarly, if the person has something placed in their left hand (and they can't see it) the sensation travels to the 'touch' region of the right hemisphere where the person 'feels' it, but because this sensation is not passed to the speaking left hemisphere, they can't say what they are holding. If the images or sensations are sent to the left hemisphere only, these effects are not observed because the left hemisphere can see, feel and speak. Occasionally, the two hemispheres can compete against each other. For example, a person may start to dress themselves with one hand, whilst the other hand tries to un-dress them. One hemisphere might start to make the person sit down whilst the other makes them stand up **(22).**

Most of us can do different tasks with our two hands at the same time, but some people with 'two brains' can perform some 'party-tricks' and get each of their hemispheres to perform some rather difficult tasks with their two hands at the same time. For example, one patient is able to right a letter with his right-hand and at the same time get his left-hand to draw a complicated picture. If you think this is easy, try it! Much of the research into split brains and our understanding of the condition was done by a neuroscientist called Roger Sperry and he was awarded the Nobel Prize for this work in 1981.

References and Further Reading:

9. The Ancestor's Tale (2016)
Dawkins & Wong
Clays

10. Principles of Neural Science (5th Edition, 2012)
Kandel, Schwartz, Jessell, Siegelbaum & Hudspeth.
Elsevier

21. Ross (1984)
Right hemisphere's role in language, affective behavior and emotion.
Trends in Neuroscience 7:3342-346.

22. Tales from both sides of the brain:
A life in Neuroscience. (2015)
Gazzaniga
Ecco Press

CHAPTER 10

Was I Dreaming?
(Sleeping and Dreaming)

Synopsis: This Chapter explains the processes of being awake, asleep, conscious and un-conscious. It also considers the need for sleep and the meaning of dreams.

I'm Awake

We are all familiar with being awake or asleep and the states or conditions of being conscious or un-conscious. Brain scientists understand certain things about these states but they are complicated and this is an area where the brain still holds many secrets and mysteries.

To be awake or conscious are the same thing to neuroscientists. They are two different words that mean the same thing. It is known that a special region in our old brain, the brain stem, 'fires-up' and this is connected to our two cerebral hemispheres which are then switched on. Our activated cerebral hemispheres then produce the state of being awake or conscious. This situation is rather like the brain stem being a 'power station' that generates electricity which is sent to buildings and houses in the cities where it is consumed. In this analogy, if the power station stops working, the buildings in the cities will also stop working due to the absence of electricity. Similarly, if this special region in the brain stem stops working, the cerebral hemispheres will not be 'switched on' and we will not become conscious that is, we will remain un-conscious. Although it is well established that the brain stem switches on the hemispheres with it's action potentials or 'sparks', it is not understood how the cerebral hemispheres can then generate the state of being awake or conscious. This subject is discussed later.

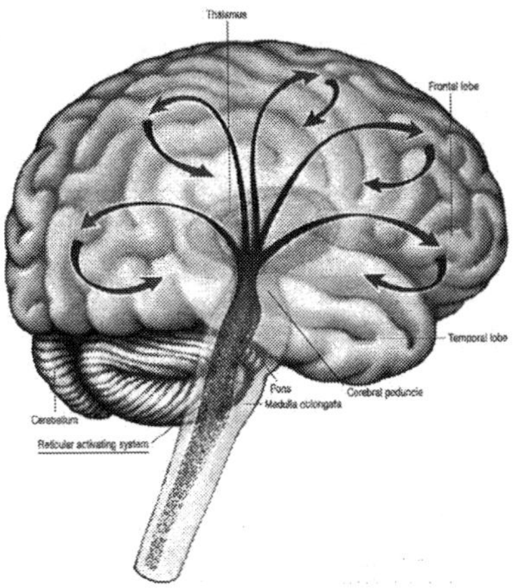

CONSCIOUSNESS

Are You Asleep?

You might think that to be asleep or un-conscious are also two different words that mean the same thing. But this is not the case. It is more complicated. It is true that if a person is not conscious they must be un-conscious; but it is possible that an un-conscious person might not be asleep. We can lose consciousness due to a variety of reasons. It is possible that the special region in the brain stem that 'fires-up' the cerebral hemispheres stops working, perhaps due to damage and if this happens the person will become un-conscious and sadly, they will never re-gain it. This is called entering a coma or comatose state. There is also the situation where a patient may be given a 'general anaesthetic' which removes consciousness making the patient un-conscious. In both these examples, it would not be correct to say the person was just 'asleep'.

Sleep is a special type of un-consciousness; It has many definitions but typically it is defined as 'A condition of body and mind which occurs naturally for several hours every night, in which the brain is inactive, the eyes closed, the muscles relaxed and consciousness is suspended'. The word 'sleep' therefore, carries the assumption that this period of being un-conscious is normal, natural and temporary that is, we will awake from it. Sadly, we may never 'awake' from a comatose state. Before looking into how we sleep, there is the fundamental question of 'Why do we and most other animals sleep at all?' That is, starting from the moment of birth, why don't we stay awake all our lives thereby removing the need to bother with this complicated process.? You could answer that we need to sleep to rest and re-charge our batteries, but I could answer that we could have been 'designed' so we don't need to ever rest.

I think the reason for sleep is linked to our spinning world where here on Earth we have a natural pre-determined cycle of about twelve hours of day-light followed by about twelve hours of night-darkness, repeatedly. We have also been 'designed' with the very special sense of vision and I'm sure we all know how difficult it is for us to do anything in perfect darkness! That is to say, vision is so important to our normal lives that we become almost useless without it. This being the case, there is some sense for us to hide away during darkness when we are useless and vulnerable. So, nature has determined that we might as well take advantage of this dark 'off-time' and so we have been programmed or hard-wired to sleep and rest at night for about eight hours **(1).**

Although our sleep / awake cycle appears linked to our natural day / night cycle, this is not entirely true. It probably started out that way, but sleep research has shown it to be more complicated. Scientists have taken subjects and put them in rooms where every single clue about the time of day has been removed. That is, there are no clocks and no sounds just no clues at all to what time of day it is or even what day or week it is! Under these conditions results have shown that everybody adopts a personal sleep / awake cycle. The subjects could choose to never sleep or to sleep for ages but everybody chooses to sleep for about eight hours and then be active and awake for about twenty hours before they choose to sleep again. When sleeping they choose it to be dark and when awake they choose to have the lights on. This shows that we all have a natural sleep / awake cycle of twenty-five to thirty hours which is slightly longer than the twenty-four hour cycle we are forced to live in.

The simple definition of sleep above says that during sleep the brain is inactive. Sleep research has shown this is not true. Sleep is not a simple 'shut-down' state that happens when the brain is switched off. Rather surprisingly, parts of the brain become very active during sleep. Therefore, during a normal night's sleep of about eight hours, our brain enter periods of about two hours when it is very active followed by a similar period when it is rather quiet. During the active periods our muscles become paralysed and our closed eyes move rapidly. This eye movement has caused this period of sleep to be called Rapid-Eye-Movement sleep or REM for short. The quiet sleep periods are called 'Deep sleep' which we find most rest full **(23).**

Sleep Difficulties

Although it is likely that sleep evolved to take advantage of our daily periods of darkness giving us a period to rest in, sleep has become an essential part of living because without it we would die! It is likely that you have missed a nights' sleep and if so, you feel terrible the next day. Rather like jet-lag. It is difficult for most people to stay awake and avoid sleep for more than three days. The world record for staying awake is about two weeks but for most of us after three days inherent biological processes in our brains take over and they shut-down consciousness forcing us to sleep! This can happen suddenly and almost secretly so we suddenly find ourselves waking up.

This shut-down process is a protective one because after about three days without sleep we experience extreme tiredness, confusion, memory lapses, hallucinations, hand tremors, headaches, bags under the eyes and increased stress hormone levels to name a few effects! If sleep is further deprived, we can experience seizures and psychosis. Many of these unpleasant effects relate to problems with brain function but problems also start to occur in our body weight, our hormone system and our cardio-vascular system and ultimately we have some kind of system shut-down that can result in death! This shows that sleep is not an option for healthy living but it has become an absolute requirement for life.

Even if we have the required amount of about eight hours of sleep a night which avoids the serious dangers of sleep deprivation, the quality of our sleep is very important to us. Most of us dream in REM sleep although missing this kind of sleep does not seem to cause any major problems in most people. But missing the other kind of sleep called slow-wave sleep or deep-sleep causes more difficulties. During this kind of sleep our body enters a maintenance routine which checks and restores deficits in our body's systems. This kind of sleep also provides proper mental rest which prepares us for the next day. Without it we continue to feel tired and become stressed and we find it hard to work, learn, create, communicate and we can have disturbed emotions. Good deep-sleep is therefore, very important to us.

As mentioned, we commonly dream during REM sleep and since most animals have REM sleep we can assume that most animals also dream! Although most people will have a few dream periods during a night's sleep, it is rare that we remember dreams. This is because dreams only enter our very short-term memory so they are usually forgotten by the time we awake. But, if we awake during a dream when it in our short-term memory, we commonly remember it the next day.

Am I Dreaming?

Why we dream remains a mystery. Neuroscientists and psychologists have different explanations. One example is the famous Austrian psychologist Sigmund Freud who lived around 1900. He proposed the process of 'Psychoanalysis' and he believed that the function of dreams was to allow us to fulfill our private wishes that we could not fulfil whilst awake. He also thought that most dreams had a sexual basis so that everything we 'saw' during a dream was coded for un-fulfilled sexual fantasies. This interpretation of dreams means that the content of a dream is very important and meaningful **(24).**

Dreams

Today many neuroscientists do not consider this to be correct. Some have proposed that during sleep our brains analyse our recent experiences and then it stores some of them in our long-term memory. This analysis and storage process they argue, produces the dream which is only a 'side-effect' of this storage process so that the content of the dream is nearly meaningless. It is certainly true that sleep-scientists have managed to stop adults from dreaming for long periods of time and this does not seem to cause them any significant harm. Therefore, it remains hotly debated whether dreams are required or not and whether their content is meaningful?

Dream research has revealed that most people do not dream wild sexual fantasies and most dreams are related to recent events. They are commonly bizarre and dis-jointed, often sad and sometimes very frightening (night mares). The dreamer may feel restrained and this might be because during dreams our limbs are actually paralysed. Although most people can sleep well at night and awake rested, this is not the case for everyone. There are many conditions that can stop a person sleeping properly and these can harm good health. There are also many conditions when a person sleeps too much which can also damage our health.

References and Further Reading:

1. Berne & Levy. Physiology (6th Edition, 2010)
   Koeppen & Stanton
   Mosby
   Elsevier

23. Neuroscience (2012)
    Purves, Augustine, Hall, LaMantia
    Sinauer

24. The Interpretation of Dreams (1997)
    Freud
    Wordsworth

CHAPTER 11

I've Had a Brain Wave!

(Brain Waves and Brain Scanning)

Synopsis: This Chapter explains that the brain normally emits regular 'waves' which can be recorded from electrodes on the scalp by an EEG machine. It also outlines MRI brain scanning and the images it can produce. Epilepsy is mentioned.

Brain Waves

It is well established that brain neurons can be in 'battery' mode during which each neuron can manufacture a small voltage called the Action Potential which is a small electrical spark of only one-tenth of a volt. Around 1900 the German physiologist Hans Berger placed several electrodes on the scalp of human subjects and he tried to record the many action potentials or sparks from the underlying brain neurons. He found that the sparks manufactured by the neurons were far too small to be detected. This was because their small voltage does not conduct very far from a neuron and certainly not up through the brain tissue, the thick bone of the skull and then the scalp to the recording electrodes. If these sparks could reach the scalp electrodes, you might expect to record a constant 'noise' from the millions of action potentials firing from the millions of neurons below. However, Berger found something rather different. To his surprise he did record some small voltages, but these were regular waves which showed that the brain was 'buzzing' at a regular frequency of about 20 Hz or waves per second.

The Electroencephalogramme (eeg)

Modern neuroscience has studied these waves and it is confirmed that the normal awake brain does buzz or emit an electrical wave of about 20 Hz which is called the beta-wave. The asleep brain emits a slower wave of about 12 Hz which is called the alpha-wave. The brain also emits other waves which appear to depend upon brain state and condition. For example, in epilepsy the signal shows many 'spikes' and in dementia the waves slow to 4 Hz. Today, these waves are called the ElectroEncephaloGramme or EEG and a modern EEG machine has an array of electrodes placed on the subjects head so that the waves can be detected from many different regions of the brain at the same time, as seen below.

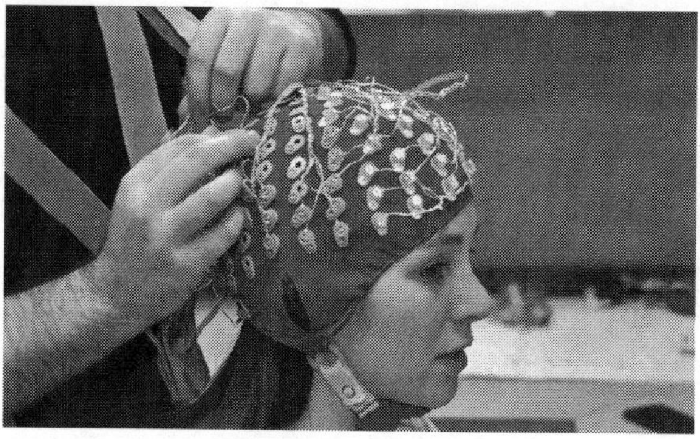

EEG Electrodes

Eeg signals are recorded and displayed on paper or on a computer screen, as shown.

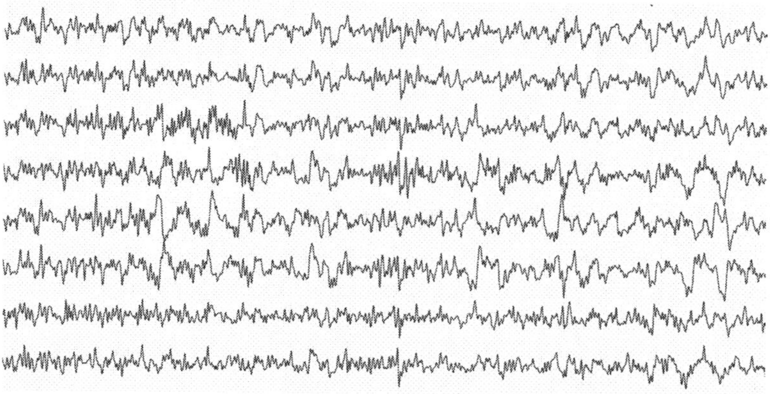

EEG Recordings

Hans Berger realised that these regular signals or waves could not be action potentials and today it is thought that they arise because many millions of neurons become aligned and synchronised and they can fire action potentials together. When they do so, the resulting electrical waves are strong enough to conduct up through the head to the recording electrodes on the scalp. However, it is still not clear exactly which parts of the brain produce these waves or how neurons become synchronised to produce them or what they are for? Recent theories think they may be connected with attention or they may be connected with the generation of consciousness. Whatever the case, recording and analysing them has become a very useful tool in modern brain clinical diagnosis **(25).**

Short Circuits!

You may have seen when there is some wrong electrical wiring you can get a 'short-circuit' which produces a flash and a bang! A similar thing can happen in the brain. There are millions of brain 'sparks' flying around the brain all the time, it never stops, and the brain wiring, that is all the neural connections have to be absolutely correct otherwise you can get a 'short circuit'. In some people this happens and the brain produces a very large 'bang' or electrical discharge. This discharge typically starts in a small part of one cerebral hemisphere and it spreads across it like a 'Mexican Wave'. This typically lasts a few seconds until the discharge decays to nothing. This is called a brain 'seizure' or a 'fit'. During the spread across the hemisphere the discharge 'excites' all the neurons along it's path. If it passes across the motor cortex (Brodmann region 4) the movement neurons are stimulated and the person will throw their limbs around in an uncontrolled way! If it passes across the touch cortex (Brodmann regions 1, 2 & 3) the 'touch' neurons will be stimulated so the person may get some strange 'feelings'. If it spreads into other areas of the brain and excites the neurons there, the person may have some peculiar experiences like 'being in a dream' or have 'religious encounters' or have 'out of body experiences' or have 'Déjà vu'. Déjà vu is French which means 'already seen' and this is the feeling that the person has 'lived these moments before'. Life is exactly repeating itself!

Our two cerebral hemispheres are connected by several large 'cables'. The largest of these is called the Corpus Callosum, which is Latin for 'Large or Tough Body'. It is about ten centimeters long and has about 25,000 'wires' or neuronal axons inside it. If the brain 'discharge' is large enough, it can spread from one hemisphere along the Corpus Callosum into the other hemisphere and if this happens, the person will lose consciousness which, of course can be very dangerous. If a person has these seizures regularly the condition is called 'Epilepsy.' Epilepsy can usually be controlled by medicines but sometimes they don't work. In these extreme cases, the person may have neurosurgery that cuts their Corpus Callosum. The benefits of this surgery is that the discharge can no longer spread to the opposite hemisphere causing loss of consciousness, which is better for the person's health. However, it also can produce some unusual effects upon the person, as mentioned earlier.

Brain Scanning

The EEG is a rather old, but still very useful, technique that can study the living brain. About fifty years ago science invented the first brain 'scanner' and the most modern of these is called Magnetic Resonance Imaging or MRI for short. An MRI scanner is a large tube that contains powerful magnets and the patient's whole body or head is positioned inside the tube, as shown. The scanner produces a very strong magnetic field which passes through the patient's whole body and as it passes through it loses some energy to each and all of the millions of atoms that make up the person's body and each atom gains a little energy. The magnetic field is then switched-off and the energy gained by all the patients' atoms is suddenly lost from them and this lost energy emerges from the body as a radio wave. This emerging radio wave is then detected by an ariel placed close to the scanner.

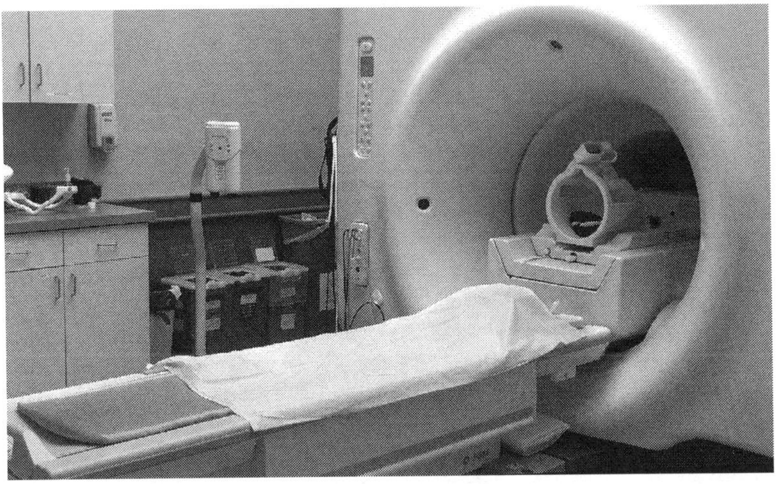

MRI Scanner

The pulse of magnetism that passes through the patient is then repeated many times which makes the atoms of the body 'buzz' or resonate; hence the name of the scanner. It turns out that different types of body tissues resonate at different frequencies and produce different radio-waves which can be recorded and analysed. This information is then used to construct a detailed image of the living brain inside the skull, as below.

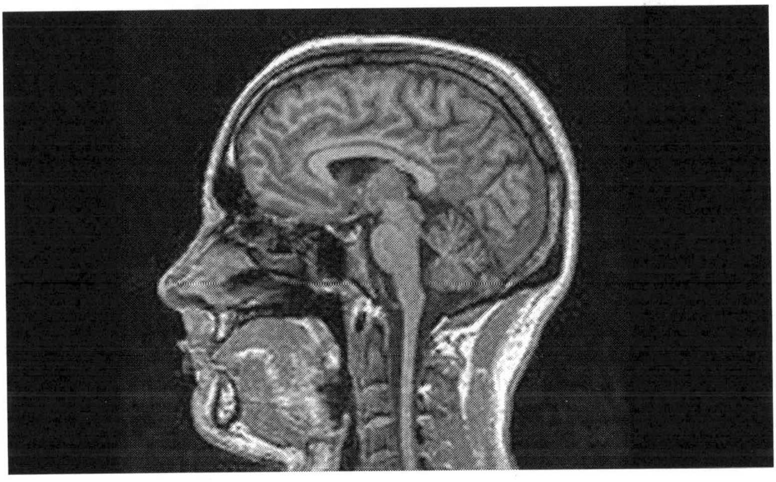

MRI Image Of The Normal Brain

Cancer cells in the brain resonate at a different frequency from normal cells so this technique can show the position of a brain tumour (white areas) which is most useful for medics.

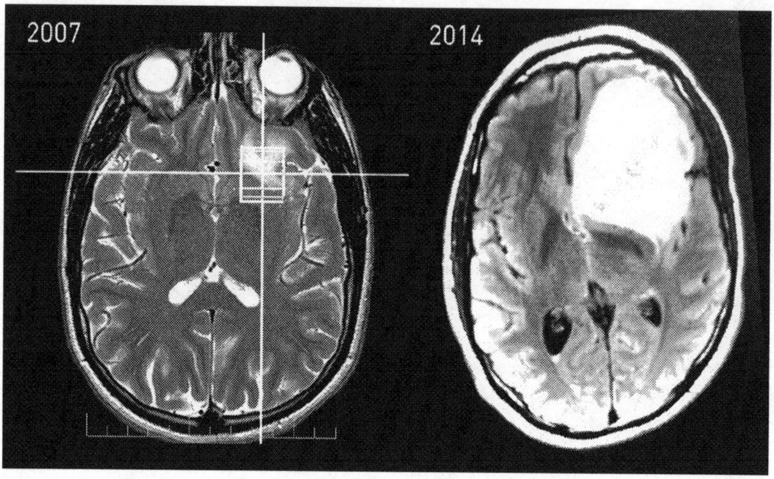

MRI Image Of Brain Showing Cancer Cells

In addition to being able to construct an image of the structure of living brain inside the skull, some types of MRI scanner can also determine which parts of the brain are functioning and which parts are not. This type of MRI scanner is called functional MRI or fMRI for short. Blood cells carry oxygen to the brain and these cells have a particular resonate frequency. When brain regions are functioning, they pull the oxygen from the blood cells which are left without oxygen and these cells have a different resonate frequency. This difference can be detected by the fMRI scanner and from this information, the scanner can create an image which shows which parts of the brain are working and which parts are not, as seen below. This kind of information has enormous clinical uses for understanding brain conditions such as Stroke when a region of the brain might have stopped working **(26).**

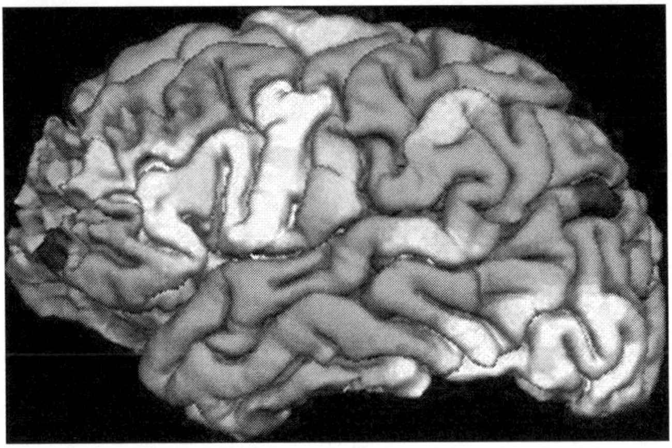

fMRI Image Showing Pale Regions Of Activity

References and Further Reading:

25. Niedermeyer's, Electroencephalography (2004)
Basic Principles, clinical applications and related fields
Schomer & Lopes Da Silva
Lippincott, Williams & Wilkins

26. Introduction to functional magnetic resonance imaging (2002)
Buxton
Cambridge Uni Press

CHAPTER 12

Protective Walls
(The Blood-Brain Barrier)

Synopsis: This chapter explains how and why the brain is surrounded by a protective 'wall'. It also explains how important nutrients are able to cross this wall and get to brain neurons.

Defensive Walls

Many groups of people throughout history have protected themselves by building a wall around themselves. An early example is 'Hadrian's Wall'. Hadrian was a Roman Emperor who built several walls at the borders of the Roman Empire to defend it against hostile invaders. The most famous stretch of his wall is close to the present border between England and Scotland which was built around 100 AD. The image shows the wall to be only a few feet high, but when it was first built is was about six meters high and three meters wide. The top of the wall had additional ramparts and there were ditches and mounds on both sides. Every mile along its length there were small 'Milecastles' and a substantial fort was constructed about every five miles. There were many guards and guarded gates along its length.

Hadrian's Wall

Later in history around 1400 the Chinese built what is now the famous 'Great Wall of China' and more recently in 1960 the German Democratic Republic built the 'Berlin Wall'. This was a guarded concrete barrier that divided the city of Berlin into Western and Eastern halves. All these protective walls were built with the purpose of protecting the builders from outside invaders and they generally worked. However, defensive walls like these also have disadvantages because they commonly stop desired trade between inhabitants on both sides of the wall. This particular problem can be solved by introducing controlled 'gates' into the wall and a famous one in the Berlin Wall was the Brandenberg Gate.

I mention protective walls because the human brain also has a protective wall around it called the Blood-Brain barrier. All tissues, organs and cells of the human body need energy to stay alive and work. This energy is supplied as oxygen, glucose and vitamins and these are delivered to our cells in our circulating blood. The small blood vessels that supply our cells are tiny tubes called capillaries. Any single blood capillary is not very long, only about one millimeter and they are joined together in an extensive branching network which ensures that every cell in the body lies close to one. Although any single capillary is not very long, we have many of them and if it was possible to join them all together to make one single tube, this thin tube would be able to stretch all the way around the Earth, making it about 25,000 miles long! All these capillaries that supply all our cells have holes in their walls which allow oxygen, glucose and vitamins to quickly pass out through the capillary walls and reach our cells beyond. When nutrients are consumed by our cells they are converted into 'waste products' and these pass back into the blood capillaries through the same holes in their walls and they are removed from the body.

I have just mentioned that all blood capillaries that supply all our cells have holes in their walls. This is true, but there is one exception; the brain. The brain does have blood capillaries, but brain capillaries don't have any holes in their walls, making them unique. This absence of holes makes it nearly impossible for anything in the blood to simply pass through a brain capillary wall and get to the brain neurons! These special 'hole-less' capillaries, therefore, form a 'wall' between the blood and the brain and this is called the 'Blood-Brain barrier' **(10)**.

Why should brain capillaries be special in this way and form a barrier? The answer is the same reason why protective walls are built namely, to protect the builders environment from disturbance and invasion. In the case of the human body, there would not be great distress if the environment of the skin was disturbed. It might be uncomfortable, but it would not be fatal. However, because the brain is the single most important control centre of the body, if the brain's environment was continually being disturbed and 'invaded' our neurons could not work properly and this could be fatal!

The concentration of oxygen, glucose, vitamins and minerals in our blood is constantly changing throughout the day. For example, the sodium levels in the blood will rise after we have eaten some salty chips and you may recall that sodium is crucial to neurons which use it to construct their important electrical 'spark' or action potential. So, the brain does not want it's sodium levels to be continually disturbed. Large fluctuations could be fatal so to reduce this possibility the brain has built its own protective wall; the Blood-Brain barrier.

This barrier does, however, also create some problems just like protective walls do. Perhaps the most obvious problem is that the presence of the barrier stops the brain's required nutrients like oxygen, glucose and vitamins from leaving the circulating blood capillaries and getting to the brain's neurons where they are required! But, just as protective walls have guarded gates, the Blood-Brain barrier also has special guarded 'gates' that can be opened to allow required nutrients through. These gates allow the brain to receive all it's required nutrients and at the same time, the wall provides a protected environment. So presence of the Blood-Brain barrier is crucial to keeping us alive! I was interested to know how many important nutrients are able to cross through the 'gates' in the Blood-Brain barrier so I undertook some research into this subject at Sunderland University, School of Pharmacy. I particularly looked at how elements like zinc manage to cross the barrier **(36).**

Barrier Difficulties

You can imagine that if the Blood-Brain barrier 'opened' following head injury for example, the brain's special and controlled environment would be disturbed and when this happens, it commonly causes brain neurons to 'fire' excessively and the person may suffer a 'fit' or brain seizure, rather like epilepsy. The presence of the barrier may also cause some difficulties for brain cancer treatment. One method of treating brain cancer is to administer 'chemotherapy' drugs to the patient. These are usually given into the patients circulating blood which carries the drugs to the brain capillaries. However, many chemotherapy drugs can not leave brain capillaries that is, they can not cross the Blood-Brain barrier because brain capillaries don't have any special guarded 'gates' in them to allow the drugs to pass through. So unfortunately, the Blood-Brain barrier can make it difficult for some chemotherapy drugs to get into the brain where they are required to work on brain cancer cells.

References and Further Reading:

10. Principles of Neural Science (5th Edition, 2012) Kandel, Schwartz, Jessell, Siegelbaum & Hudspeth. Elsevier

36. Pullen, Franklin & Hall (1991). 65-Zinc uptake from blood into brain. Journal of Neurochemistry 56, 485 – 489.

CHAPTER 13

Think About It
(Thinking & Beyond)

Synopsis: This Chapter explains how the brain thinks and makes us self-aware. The area of mental illnesses is covered and the subjects of brain, mind, life and death plus our place in the universe are considered.

Thinking & Self Awareness

It has been mentioned that each of our senses are continually sending information to the brain so we can see, hear, smell, taste and touch and this information arrives at the same time. To start with, each of these sensations arrives at its own destination site in the brain but somehow the brain is able to blend all these sensations and experiences together to form our total conscious experience of the world from second to second. Neuroscience does not know how the brain does this. It also does not know how we 'think' or how we are able to make our thoughts become commands that allow us to move and do things. Science is not even completely sure how the brain is able to generate consciousness in the first place although some part of the process is understood. It is established that our brain stem switches on our cerebral hemispheres, which then somehow generate the state of wakefulness **(27).**

Although it is known that the awake brain has many neurons firing, this alone cannot explain consciousness. It is possible that consciousness is created by the synchronised firing of millions of neurons as first recorded with the eeg. However, some modern theories of consciousness are entering into the realms of 'quantum mechanics'. These theories propose that quantum processes are taking place inside our brains and this is a level of science more subtle than electrical 'sparks' alone. If true, it could mean that our brains can enter into other dimensions and space-time continuums which are not yet recognised by science **(28).** Whilst being conscious we are thinking and we are self-aware. That is, we know we exist and we know we are mortal and as the French Philosopher Rene Descartes said in 1637, 'Je pense, donc je suis'. Translated to 'I think, therefore I am'. Meaning, I am aware that I am thinking, so I must be alive and exist as a person!

The process of thinking remains a secret and mystery of the brain and recent brain research has shown that we all might be 'living in the past'! For example, everyone blinks their eyes for about one-third of a second and everybody blinks about once every ten seconds all the time. This means we are all continually closing our eyes and even though its only for a fraction of a second, we don't experience these un-sighted periods. Sight for us is constant and this is because our brains are constantly editing-out the blinks making us see in an un-interrupted stream. We are not aware of the removal of the 'blank moments' which means we are seeing the world following a short delay so we are living a little in the past.

Another example of our brains' secret intervention in our world are eye 'saccades'. Saccade is French meaning 'jerk' and when we look at the world, our eyes do not look at a scene with fixed steadiness, instead they jump around making jerky 'saccadic' movements, stopping many times and then moving on as fast as they can between each stop. These movements allow our eyes to look at different parts of the visual scene and build-up a three-dimensional 'where' map'. But the brain does not release all these jerky movements to us because if it did our vision would be a constant 'blur'! Instead it intervenes and removes all the blurs and only releases it's edited version allowing us to see smoothly! Again, this means we only become conscious or aware of visual scenes after our brain has secretly processed them first! Sometimes, it is possible to detect this process happening with the 'clock experiment'. You may have noticed that if you suddenly look at the second-hand of a clock, it seems to 'freeze' still for a moment before it ticks-on to the next second. This 'freezing' moment is due to the time delay caused by our brain collecting all the 'saccadic information' from both eyes before it passes on the final image to us.

Equally surprising is that using brain analysis machines like eeg and fMRI, neuroscientists have recorded neuron activity in the brain when we make a decision to move. They found that from the moment we first decide to move, there is a delay before the movement is actually made. This delay is the time it takes the brain to work out what has to be done, write a movement 'programme', recruit all the required neurons and then finally execute the move. There are time delays in vision already mentioned between the brain receiving an image and us becoming aware of it and these delays are fractions of a second. However, for movements, the time delays are thought to be several whole seconds long! Similarly, these delays are never experienced by us in life because our movements seem to occur the very instant we decide to do them. These results have produced the theory that when we make a decision in our minds to move, we don't actually become aware or conscious of our own decision for a few seconds and our own brain is able to secretly hide this delay from us, rather like it does for aspects of vision, as mentioned earlier. So, if the brain can hide from us our blinks, our jerky eye movements and our movement decisions, might it also be doing the same for other sensations, thoughts and decisions! **(29).**

Whilst we guess that most animals alive like cats and dogs also think, few have insight into life and their existence. We do, some monkeys and apes do and possibly whales and elephants, but no other creatures are thought to know they exist as living, mortal beings. One way to see if animals are self-aware is called the mirror-test. If you hold a mirror up to the face of an adult human, the person should immediately recognise the image in the mirror is themselves. Babies will pass this test once they are one or two years old. If you hold the same mirror to the face of a dog, it will not recognise the image in the mirror as itself and probably not even recognise it to be a dog. Some dogs might perhaps recognise the image as a dog and look behind the mirror for the dog and bark at it, but they don't have the brain power to understand that the image is actually a reflection of themselves. They don't have that degree of self-awareness. Chimpanzees and some great apes can see themselves in a mirror and to support this, if you put a coloured sticker on the nose of one without them knowing or put a hat on their head without them knowing and present a mirror to them, they will see the unusual sticker or hat on themselves and remove them. This shows that they can see and recognise themselves and have enough insight to be self-aware.

Mirror Test

Associated with our self-awareness is the knowledge that we are all born, live and die but most animals do not know this. If you happen to pass a field full of sheep or cows and one of them is lying dead on the ground, the rest of the herd will not know its dead. They might know it's not moving and perhaps it smells different, but they will not understand it is 'no more' and grieve its loss. Humans can grieve and understand mortality. Behavioral scientists think that some monkeys, apes, whales and elephants can also grieve the loss of loved-ones because they have brains big enough to have these complicated feelings. Smaller brains probably can't manufacture such complicated emotions.

The original work of Brodmann and Broca and Wernicke and others have shown that we know what some parts of the brain do. We also know something about the areas at the front of the hemispheres called the frontal lobes. These are thought to be responsible for drive, motivation, personality and intellect which combine to produce our greatest and most sophisticated thoughts and processes that give us our true human qualities. Years ago when some patients in hospitals who had mental illness could not be controlled, they might have had their frontal lobes surgically removed because this was found to pacify them. This surgery was called a 'Frontal Lobotomy'. It often worked, but it is not used to-day. I would rather have 'A bottle in front of me'!

Neuroscience has also discovered that there are other large areas of the cerebral hemispheres involved with 'understanding' the sensations that are received. As mentioned with vision, it is one thing to be able to 'see' shapes, but it is another league of difficulty to be able to put it all together and see a whole landscape like the Hay Wain. The same applies to hearing. It is one thing to 'hear' something, but it's much more involved and complicated to be able to understand what you are hearing. Similarly for smell, touch and taste, so that in addition to receiving sensations, we have to understand them so comprehension of our sensations might be what half the brain does!

Mental Difficulties

In many of the sensory difficulties mentioned so far, problems can be caused by neurons in particular regions of the brain being destroyed so their function is lost. People will have difficulty with speech following damage to Broca's Area, as an example. Mental conditions or illnesses or disorders are more complicated because there may not be any damage observable in the brain. This means that these illnesses are likely to be caused by neurons not working properly rather than regions having been destroyed. Neurons not working properly can often be made to start working correctly again and this is where counselling and medicines step-in to help. As mentioned, if you have a complicated and sophisticated network or system, like the production of our mental state or production of our 'mind' and part of this network fails, it might all 'crash' and totally stop working, but most likely it will continue working but with some strange abnormalities. This is the case with mental health.

Anxiety is a normal human response to a threat and is usually required to help us to overcome a difficult situation. However, excessive anxiety is not good because it might be too intense or because it happens in situations that do not really present danger. Its' symptoms range from a heightened sense of awareness to a deep fear of impending disaster and even impending death! Anxiety disorders are the most common psychiatric disorders, found in about one quarter of the adult population. Key symptoms are fear, arousal, restlessness, sweating, racing heart, dry mouth and the desire to escape. It is classified as an illness by itself and also a component part of Panic disorder, Phobias, Obsessive-Compulsive Disorder and Post-Traumatic Stress Disorder. There is not any known brain damage in anxiety although abnormal function of a brain structure called the Amygdala may be involved. This structure lies deep within the brain which acquired its name because it looks a little like an almond nut (Amygdala in Greek). The Amygdala is involved in the formation and storage of memories associated with emotional events, in particular fear. Thus, over activity in this structure is likely to contribute to anxiety disorder. Anxiety can be treated with counselling or anti-anxiety medicines.

Depression is a 'mood disorder' and it is commonly more severe than anxiety. It is usually a life-long disease and is twice as common in women as in men. The depressed person experiences a general slowing down and a loss of energy and enthusiasm for life. There may be delusions of other illnesses and an expectation of severe punishment. It may be life-threatening because about one quarter of patients attempt suicide. The depressed person has a negative attitude, loss of interest in life and feels there is no possibility of improvement. Such persons commonly avoid eye contact and have slow movements and slow responses to questions. Commonly there is weight loss, early morning awakening, constipation and loss of libido.

There is not any known brain damage in depression although abnormal function of areas of the cortex is thought likely. The absence of brain damage plus the looseness of its symptoms has led some medics to dismiss it as a true illness. It is treated with counselling, anti-depressant medicines and perhaps ECT. ECT stands for Electro Convulsive Therapy and this involves the placement of electrodes on the persons head and whilst they are anaesthetised, high voltage is applied to the electrodes which makes the person have a brain seizure / fit! Eight treatments are commonly given at two-day intervals over a period of four weeks. It is not clear why this extreme treatment works but it does appear to help in many cases.

Schizophrenia means 'splitting of the mind' and it is classified as serious Psychosis. Psychotic individuals exhibit impairment of intellectual function and commonly have feelings of persecution, thought control, extreme fear, agitation, emotional lability, social withdrawal and apathy. Commonly these persons have visual hallucinations and hear voices. The voices are usually disturbing and they tell the person how bad they are or they demand them to do bad deeds! Such persons also have agitation, suspiciousness and paranoia. Modern theories of Schizophrenia think that the voices they hear could be their own thoughts but they don't realise this and consider them to be external. Altogether, there is a serious dis-connect from reality. There is some brain damage in schizophrenia although it is thought that the condition is largely due to the neural connections between certain brain structures such as the hippocampus and frontal lobes being quite wrong. This mis-wiring may cause the person to experience the world in a rather bizarre way. Treatment of the condition includes counselling and anti-psychotic medicines **(30).**

Dementia means 'without mind' and it is a mental illness that has recognised brain damage unlike many other mental illnesses. Dementia results from the degeneration of the neurons in various parts of the brain with resultant loss of their function. The most common type of dementia is Alzheimer's Disease which was named after the German pathologist Alois Alzheimer who first described it as a formal disease around 1900. It is most common in women over eighty years old. An early symptom is the loss of short-term memory followed by the loss of long-term memory and then a general intellectual decline. There are many tests used clinically to evaluate the mental loss and below are some the questions commonly asked.

MENTAL TEST

- Name
- Age
- Time
- Remember the following address;

  Mr. S Holmes, 221B, Baker Street
- Day of week
- Month
- Year
- Place of interview
- Name two people present
- Birthday
- School attended
- Occupation
- First name of spouse
- Name of the Queen
- Name of the Prime Minister
- Count 1 to 20
- Recall the address

There are about twenty questions in this test and most of us would answer all of them correctly. But a person with severe Alzheimer's Disease may only score 1 or 2. Sadly, people with the advanced illness cannot remember or recognise their family members and they may not even recognise themselves in a mirror. They may not be able to dress, wash or feed themselves so they need full-time care. They might also see the world in a rather odd way as illustrated by the famous case of the 'Man who mistook his wife for a hat' **(31).**

There is known to be brain damage in dementia and I was fortunate enough to do some research into the causes of this illness at Brown University in America, which is an 'Ivy League' University **(32).** By the way, the Ivy League Universities are the oldest Universities in America and some of them are old enough to have Ivy growing over their walls; hence the name. I continued to investigate why parts of the brain degenerate in Alzheimer's Disease and whether exposure to metals like aluminium are involved in the cause of this disease. This research was done at the Medical Research Council laboratories in Newcastle **(33)**.

Brain and Mind

I have discussed the human brain as a collection of millions of sparking neurons, but psychologists also discuss the 'mind'. There are many Professors of Psychology working in Universities around the World today who argue the 'Mind' is an entity quite separate from the physical brain and in some ways this view is correct. We do not simply experience electrical sparks in our brains, rather, we experience thoughts and emotions and you could argue these are the 'mind'. However, I'm not sure how separate the mind can be from the brain because I don't see how it can exist without a brain, so I consider the mind to be a clever creation of the physical brain. I could be wrong, but science has learned a lot since Roman times and some parts of the mysteries of life can now be explained, but not all!

Brain Limits!

Although our body's and brains are capable of sensing much about our outside world, there remains much about the world that our brains cannot sense. For example, at this very moment whilst you are sitting and reading this book there are many radio-waves flying straight through you, yet you can't feel them because our body does not have any method of sensing them. Because you can't sense them, you might be tempted to say they don't exist, but science would prove you wrong. Similarly, there might be many other dimensions and even other universes around us that we can't detect, and you might be wrong to say they don't exist simply because you can't detect them.

If modern neuroscience is correct, you could argue that everything we experience outside us that is, matter, anti-matter, dark matter, gravity, atoms, strings, time and space are all just a series of brain sparks. You could also argue that everything we experience inside us like our thoughts and emotions are also just a series of brain sparks. This being the case, there is an argument that actually life, the universe and everything doesn't exist as such. It is all just a clever creation of the brain which has managed to put everything together in a skilled and subtle way that fools us into believing that the outside world is really there! I think this is a rather extreme argument, but it does provide us with something to think about.

Whatever is the answer, it could be that the human brain at its present stage of evolution is not complicated, sophisticated or clever enough to understand the nature of our infinitely large universe or our infinitely small atoms. Consider this; suppose you take a Goldfish and place it deep within the Atlantic Ocean half way between England and America and you allow it to swim away in any direction it chooses. Whichever way it goes, up-down, left-right or forwards-backwards, it will never reach the end of the ocean because this is just too far for it to swim in its lifetime. This means that from the Goldfish's view point, no matter which way it swims, the Ocean is infinite and endless.

GOLDFISH

Of course, we know the Atlantic Ocean has boundaries because our brains are clever enough to understand the dimensions of the world we live in. But a Goldfish can't understand this and no matter how much you try to 'explain' this to the Goldfish it will never understand because it doesn't have the brain capacity to do so! Similarly, when we venture out into space, no matter which way we go we find it endless and infinite in every direction. And when we study atoms, no matter which atomic particle we look at, we always find it is composed of even smaller particles. Ad infinitum! Are we then perhaps Goldfish and we just can't grasp the advanced concepts and dimensions of our universe and no matter how much it is explained to us we will never grasp it?

References and Further Reading:

17. Neuropathology:
A Reference Text of CNS Pathology (3rd Edition, 2012)
Ellison, Love, Maria, Harding, Vintners, Brandner, Yong. Mosby

27. The Psychology of consciousness (1992)
Farthing
Prentice Hall

28. Fashion, Faith and Fantasy in the new physics of the universe (2016)
Penrose. Princetown Uni Press

29. Soon, Brass, Heinze & Haynes (2008)
Unconscious determinants of free decisions in the human brain.
Nature Neuroscience 11, 543 - 545

30. Diagnostic and Statistical Manual of Mental Disorders (5th Edition, 2013)
American Psychiatric Association
American Psychiatric Publishing
Nature Neuroscience

31. The man who mistook his wife for a hat (2011)
Sacks
Picador

32. Pullen, DePasquale, Cserr (1987).
Bulk flow of cerebrospinal fluid into brain in response to acute hyperosmolarity. American Journal of Physiology 253, F538 - 545.

33. Pullen, Candy, Morris, Keith, Taylor, Edwardson (1990). 67-Gallium as a potential marker for aluminium transport in brain: implications for Alzheimer's Disease. Journal of Neurochemistry 55, 485 - 489.

CHAPTER 14

Hasta La Vista, Baby!

(Artificial Intelligence)

Synopsis: This Chapter discusses differences and similarities between neurons and computer-chips. Computer-chips can process data much faster than brains, but brains have greater overall processing capability. Artificial intelligence is also discussed.

Data Processing

I have mentioned that brain neurons can go into 'computer-chip' mode. That is, a neuron can receive thousands of signals every second, count them, analyse them, process them and then decide to do something, like send out it's own signal or perhaps do nothing. Neurons have their own language which is rather like Morse code and modern computers with silicon-chips also use a kind-of Morse code which is called 'Machine code'. In machine code, a computer chip can manufacture a 'dot' which is a tiny electrical voltage or 'spark' and it can also make a 'dash' which is the absence of a spark, just like the brain. Or one can think of this machine code as a series of ones (1s) and zeros (0s). Importantly, from this simple machine code complicated signals and instructions can be sent and received in computers just like brain neurons. Sending, receiving and de-coding signals is called 'data processing'.

Modern computer-chips can perform data processing much faster than the human brain can. This is because human brain neurons can only generate about two hundred sparks per second where as a modern silicon-chip can generate about two hundred-million sparks per second. This means for example, if a mathematician sat down and worked on a complex calculation and it took them twelve hours to complete it, the same calculation could be finished by a computer in just one second! This tremendous speed of mathematics makes computers able to control all sorts of processes that humans can't, like fly an aircraft blind and predict the global weather **(34).** Although computer-chips work really fast, the number of chips that can connect and talk to each other is limited. In contrast, the human brain has millions of neurons and although each one processes data rather slowly, each one has enormous connectivity. This means that on balance, the total amount of processing the brain is capable of is millions of times greater than the biggest and fastest computer on Earth.

This fast operation of computer-chips has led to the notion of Artificial Intelligence where computers, machines or robots become 'intelligent'. A famous test of computer intelligence is called the 'Turin' test which was devised by Alan Turing in the 1950s. Alan Turin is most famous for being the British mathematician who cracked the 'Enigma Code'. In the Second World War the Germany military sent messages to their troops in a code to stop the British forces understanding them. The German code was made using an Enigma machine. Turin managed to crack the code which led him to work closely with computers and after the War he devised the Turin Test. In his test, a person sits in front of a computer and they ask the computer any questions they like by typing them into the key board. The computer gives an answer on screen. This answer might have been produced by the computer although it might have been generated by a living person who is hidden from the questioner who can also type on-screen. To pass the test, the computer-generated answers must be good enough to convince the interrogator that it is the hidden person who is answering the questions and not the computer **(35).** Very few if any, computers today have passed this test.

Computers that can answer questions are often called 'Chatterbots' and many can be found on the Internet. But if you talk with one it doesn't take long before it becomes obvious that you are talking to a computer, not a person. They don't pass the Turin Test. A common technique used by chatterbots is to answer your question with it's own question. For example, if you ask 'Do you like swimming'? It will answer 'Why are you asking me this? If you say, 'Because I would like to know if you mind getting wet'. It will answer 'Do you?' If you say 'I'm fine getting wet, but wouldn't getting wet harm you? It will answer 'Why should it?........ and so on..........

Even though computers have very fast processing speed the most advanced computers remain rather limited. We do have computers that can 'see' well enough to guide vehicles around warehouses, build cars in factories and in the near future computer-controlled cars might be able to 'see' traffic well enough to be able to drive us around. But even the most advanced computers or robots of today can only just walk and they certainly can't generate anything like our complex senses of vision, hearing, touch, taste, smell, movement or language and are they really 'intelligent'?

Robots!

A computer or robot does not understand anything it is doing or seeing or hearing and of course they don't have emotions, self-awareness or consciousness. Robots don't 'think' and can't 'love' and it's hard to think of what they would be conscious of? I am prepared to accept that in the future, technology will advance to the point where robots will pass the Turin Test and they may even look, speak and move like real people. It might be made even more difficult to 'spot' them if they are programmed to simulate our emotions. For example, they may say 'ouch' if you kick them and they could smile and say 'I love you'. Even so, I find it hard to imagine that a dead, metal machine could ever have the qualities of a living, moral person made of flesh and bones.

References and Further Reading:

34. Universal Artificial Intelligence (2005)
Hutter
Springer

35. The Emperor's New Mind:
Concerning Computers, minds and the laws of Physics (1999)
Penrose
OUP

REFERENCES:

1. Berne & Levy. Physiology (6th Edition, 2010)
   Koeppen & Stanton
   Elsevier
   ISBN-10: 032307362X
   ISBN-13: 978-0323073622

2. The Evolution of the Microscope (2014)
   Bradbury
   Elsevier
   ISBN-10: 1483131904
   ISBN-13: 978-1483131900

3. Danielli & Davson (1935).
   A contribution to the theory of permeability of thin films.
   Journal of Cellular and Comparative Physiology. 5 (4): 495–508.

4. Singer & Nicolson (1972).
   The fluid mosaic model of the structure of cell membranes.
   Science 175 : 720-31

5. Physiology of the Cerebrospinal Fluid (1967)
   H. Davson
   J & A Churchill
   ASIN: B0000CNE7W

6. The Double Helix (2010)
   Watson
   W & N
   ISBN-10: 075382843X
   ISBN-13: 978-0753828434

7. Gray's Anatomy (41st Edition, 2015)
The Anatomical Basis of Clinical Practice
Stradring
Elsevier
ISBN-10: 0702052302
ISBN-13: 978-0702052309

8. Pullen (1985).
Effect of ventricular tonicity on CSF production.
Journal of Physiology 362, 273-283.

9. The Ancestor's Tale (2016)
Dawkins & Wong
Clays
ISBN-10: 1474606458
ISBN-13: 978-1474606455

10. Principles of Neural Science (5th Edition, 2012)
Kandel, Schwartz, Jessell, Siegelbaum & Hudspeth.
Elsevier
ISBN-10: 0071390111
ISBN-13: 978-0071390118

11. Benjamin Franklin: An American Life (2003)
Isaacson
Simon & Schuster
ISBN-10: 0684807610
ISBN-13: 978-0684807614

12. Golgi: A Biography of the Founder of Modern Neuroscience
(2010) Mazzarello. Oxford Uni Press
ISBN-10: 0195337840
ISBN-13: 978-0195337846

13. Hodgkin and Huxley (1952). A quantitative description of membrane current and its application to conduction and excitation in nerve . The Journal of Physiology. 117 (4): 500–544.

14. Molecular Mechanisms of neurotransmitter release (2008)
Wang
Human Press
ISBN-10: 193411538X
ISBN-13: 9781934115381

15. Brodmann (1909)
Vergleichende Lokalisationslehre der Grosshirnrinde
J A Barth

16. Neuropathology: A Reference Text of CNS Pathology (3rd Edition, 2012)
Ellison, Love, Maria, Harding, Vintners, Brandner, Yong.
Mosby
ISBN-10: 0723435154
ISBN-13: 978-0723435150

17. Symon, Branston, Chikovani, Pullen (1979). Ischaemic brain oedema following middle cerebral artery occulsion: Relationship between regional water content and blood flow at 1 to 2 hr. Stroke 10, 184-191.

18. Losing Touch: A man without his body (2016)
Cole
OUP Oxford
ISBN-10: 0198778872
ISBN-13: 978-0198778875

19. Eye, Brain and Vision (1995)
Hubel
Scientific American Labs
New York
ASIN: B00AEWDVUA

20. Melzak & Wall (1965)
Pain Mechanisms: A New Theory
Science, 150, 971-979

21. Ross (1984)
Right hemisphere's role in language, affective behavior and emotion.
Trends in Neuroscience 7:3342-346.

22. Tales from both sides of the brain: A life in Neuroscience. (2015)
Gazzaniga
Ecco Press
ISBN-10: 0062228854
ISBN-13: 978-0062228857

23. Neuroscience (2012)
Purves, Augustine, Hall, LaMantia
Sinauer
ISBN-10: 0878939679
ISBN-13: 978-0878939671

24. The Interpretation of Dreams (1997)
Freud
Wordsworth
ISBN-10: 1853264849
ISBN-13: 978-1853264849

25. Niedermeyer's, Electroencephalography (2004)
Basic Principles, clinical applications and related fields
Schomer & Lopes Da Silva
Lippincott, Williams & Wilkins
ISBN-10: 1451153155
ISBN-13: 9781451153156

26. Introduction to functional magnetic resonance imaging (2002)
Buxton
Cambridge Uni Press
ISBN-10: 0521899958
ISBN-13: 978-0521899956

27. The Psychology of Consciousness (1992)
Farthing
Prentice Hall
ISBN-10: 0137286686
ISBN-13: 978-0137286683

28. Fashion, Faith and Fantasy in the new physics of the universe (2016)
Penrose
Princetown Uni Press
ISBN-10: 0691178534
ISBN-13: 978-0691178530

29. Soon, Brass, Heinze & Haynes (2008)
Unconscious determinants of free decisions in the human brain.
Nature Neuroscience 11, 543 - 545

30. Diagnostic and Statistical Manual of Mental Disorders (5th Edition, 2013)
American Psychiatric Association
American Psychiatric Publishing
ISBN-10: 0890425558
ISBN-13: 978-0890425558

31. The man who mistook his wife for a Hat (2011)
Sacks
Picador
ISBN-10: 0330523627
ISBN-13: 978-0330523622

32. Pullen, DePasquale, Cserr (1987).
Bulk flow of cerebrospinal fluid into brain in response to acute hyperosmolarity. American Journal of Physiology 253, F538 - 545

33. Pullen, Candy, Morris, Keith, Taylor, Edwardson (1990). 67-Gallium as a potential marker for aluminium transport in brain: implications for Alzheimer's Disease. Journal of Neurochemistry 55, 485 - 489.

34. Universal Artificial Intelligence (2005)
Hutter
Springer
ISBN-10: 3540221395
ISBN-13: 978-3540221395

35. The Emperor's New Mind: Concerning Computers, minds and the laws of Physics (1999)
Penrose
OUP
ISBN-10: 0192861980
ISBN-13: 978-0192861986

36. Pullen, Franklin & Hall (1991).
65-Zinc uptake from blood into brain. Journal of Neurochemistry 56, 485 – 489.

Made in the USA
Monee, IL
09 November 2026

68263307R00068